AN INTRODUCTION TO THE JOSEPHSON EFFECTS

Other titles in this series

TL/EE/1 LINEAR ELECTRIC MOTORS
E R Laithwaite

Other Electrical Engineering titles available

EE/1 THRESHOLD LOGIC
S L Hurst

EE/2 THE MAGNETRON OSCILLATOR
E Kettlewell

EE/3 MATERIAL FOR THE GUNN EFFECT
J W Orton

EE/4 MICROCIRCUIT LEARNING COMPUTER
I Aleksander

All published by Mills and Boon Limited

M & B Technical Library TL/EE/2

General Editor: J Gordon Cook, PhD, FRIC

An Introduction to the Josephson Effects

B W Petley, PhD, MInstP

Division of Quantum Metrology,
National Physical Laboratory

Mills & Boon Limited

London

First published in Great Britain 1971
by Mills & Boon Limited, 17–19 Foley Street,
London W1A 1DR

ISBN 0.263.51705.5

Made in Great Britain at the Pitman Press, Bath

CONTENTS

1. Introduction

A. THE JOSEPHSON EFFECTS

The Josephson effects initially described the flow, by tunnelling, of a superconducting current or supercurrent between two superconductors which were separated by an insulating barrier.[1] The effects have since been found to be of wider application, and they are also observed if the two superconductors are weakly connected, for example by a superconducting bridge or by a semi-conducting or normally conducting barrier. Analogous effects can also occur when two samples of superfluid liquid helium II are weakly linked, for example by a fine orifice.

Josephson's (1962) predictions were both startling and initially controversial, but they were soon verified experimentally. Confirmation of the Josephson effects was undoubtedly speeded by the existence at that time of several groups of experimenters in the United States who were studying normal electron tunnelling between superconductors. The necessary apparatus and experimental techniques were therefore available. Indeed, Anderson[2] and others have suggested that Josephson effects had already been observed but dismissed as experimental artifacts.

Josephson's prediction that a superconducting current could tunnel through the barrier was difficult to verify unambiguously, since an imperfectly-made barrier with a fine, short-circuiting link could simulate the same effect. However, it was predicted that in the presence of a magnetic field the tunnelling current would depend on the magnetic flux through the barrier, and this unique property was observed by Anderson and Rowell[3] and Rowell.[4]

Josephson also predicted that an alternating supercurrent could tunnel through the barrier. The existence of this a.c. supercurrent was inferred by the presence of supercurrent steps that appear in the barrier direct current-voltage characteristic both in the presence of microwave radiation, by Shapiro,[5] and in the absence of applied radiation, from the effects of self-resonant frequencies of the barrier, by Fiske.[6] The existence of the alternating super-current was later confirmed more directly by the detection of the radiation emitted when there was a direct voltage across the barrier, by Yanson, *et al.*,[7] and by the detection of the Josephson plasma resonance radiation which is emitted when there is no voltage across the barrier, by Dahm, *et al.*[8]

Figure 1 illustrates the type of cryostat arrangement employed by the group at the University of Pennsylvania for detection of the a.c. Josephson effects in emission and absorption.

The Josephson effects in superconductivity are now fairly easy to demonstrate experimentally, and a normal laboratory liquid helium storage dewar will suffice for the cryostat. Descriptions of apparatus suitable for student demonstrations have been given by Richards, *et al.*,[9] Collings and Gordon[10] and Petley and Morris.[11]

Josephson effects were also expected in liquid helium II, and Anderson[12] discussed several possible ways of observing them experimentally. The first observations of the Josephson effects in Helium II were reported by Richards and Anderson,[13] and their experiments were repeated with greater stability by Khorana and Chandre-skhar[14] and Khorana.[15]

It is beyond the scope of a short monograph to give a detailed theoretical discussion of the Josephson effects, and the reader is referred to the excellent papers by Joseph-son,[1,16,17,18] Anderson,[2,12,19] Feynman[20] and Bloch.[21] Russian theoretical work is covered in Reference

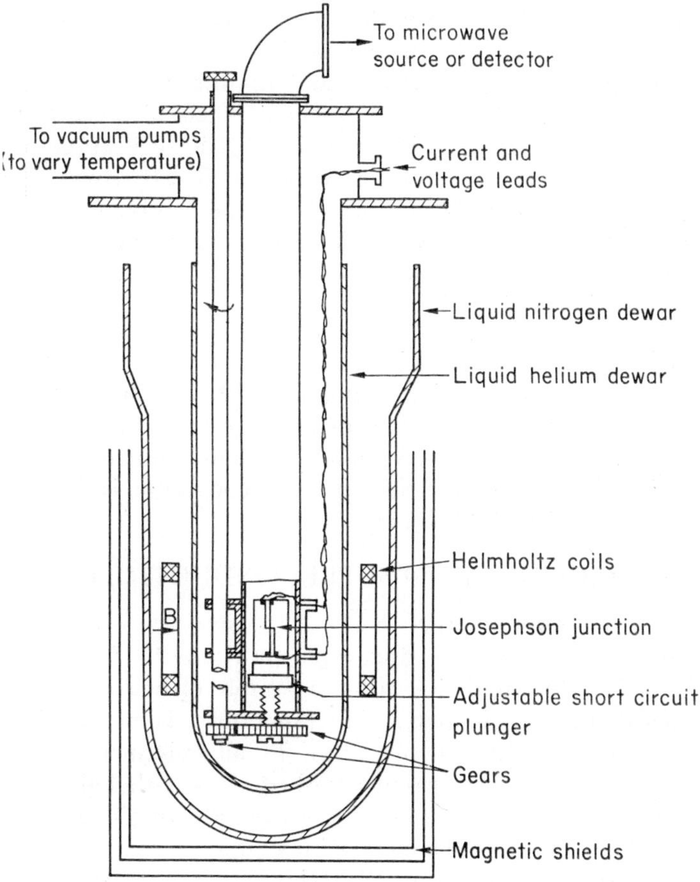

Fig. 1 Example of apparatus used at the University of Pennsylvania to study the a.c. Josephson effect by emission and absorption.

114. Sufficient detail has however been given here to show how the various Josephson effects arise.

The Josephson effects are intimately linked with the continuing development of superconductivity and superfluidity, and these are therefore discussed in the next section in order to introduce the necessary concepts, particularly the theory of Bardeen, Cooper and Schreiffer.(22)

These subjects are considered in greater detail elsewhere, for example in the books by Lynton[23] and Rose-Innes and Rhoderick[23] and a number of articles in Contemporary Physics.[24] Bardeen[25] has also given a very interesting historical summary. The collection of articles by various authors in the book edited by Parks,[26] not only gives an overall view of superconductivity at mid-1968, but contains particularly relevant articles by Mercereau and Josephson, together with Anderson's equally interesting historical survey.

It is unfortunate that physics at present is beset by two systems of units; in this monograph, SI units have been employed throughout. The many magnetic effects have been expressed as far as is possible in terms of flux densities rather than magnetic fields, since the conversion is simply 1 Gauss $= 10^{-4} T$ (Wb/m^2).

B. CONCEPTS FROM SUPERCONDUCTIVITY

(i) SUPERCONDUCTIVITY

Superconductivity was discovered by Kamerlingh Onnes in 1911. He found that when a sample of mercury was cooled below a characteristic temperature, known as the transition or critical temperature T_c, all resistance to the flow of an electric current vanished. It was thought for many years that a metal in the superconducting state was merely an example of a metal with an infinite conductivity (or zero resistivity). In 1933, however, Meissner and Ochsenfeld discovered that a magnetic field was excluded from a superconductor whether it was applied before or after the metal was cooled below the transition temperature. Furthermore, the state with the flux excluded was the thermodynamically stable state. The Meissner effect showed that superconductivity was more than simply a case of infinite conductivity, for the flux should have been frozen in or trapped when the magnetic field was applied before the metal was cooled through the transition temperature.

For many years, attempts were made to describe superconductivity theoretically. Phenomenological theories such as the Casimir-Gorter two fluid model (1933) and the London theory for the electromagnetic properties (1935) were developed to describe various aspects, while thermodynamics was also applied successfully, for example to relate critical fields with heat capacities.

Considerable progress was made towards describing experimentally observed properties following London's important proposal that the Meissner effect and other superconducting properties—for example, perfect conductivity—could be explained if superconductivity was a macroscopic manifestation of quantum effects. Thus, the property of superconductivity could be considered as being a state having a long-range order, so that the motions of a substantial fraction of the electrons throughout the entire bulk of the superconductor were correlated. These and other ideas were further clarified in London's book[27] and by Ginzburg and Landau's[28] very important generalisation of the London theory. They introduced a complex order parameter with amplitude and phase to describe superfluid flow, while Pippard (1953) made a non-local revision of the London electrodynamics.

In 1950, Frohlich had made an important proposal that superconductivity was a result of the interactions of electrons and phonons, the latter being the quanta of lattice vibrations. His theory gained substance when two groups independently discovered that the transition temperature depended on isotopic mass as predicted, indicating that the motion of the ions in the metal was involved with the property of superconductivity. Theoretical physicists considered that bound pairs of electrons (now known as Cooper-pairs) might be involved in superconductivity and Cooper (1956) showed that electrons outside the normal Fermi-sea could form a bound pair, in the presence of an attractive force, no matter how weak the interaction. Frohlich, and Bardeen and Pines had already shown that an attractive interaction of this type could arise from an exchange of virtual phonons.

(ii) THE B.C.S. THEORY AND THE ENERGY GAP

(a) *The B.C.S. Theory*

All the ideas discussed above were brought together in the beautiful theory of Bardeen, Cooper and Schreiffer.[22] In their theory they showed that London's macroscopic effect could be explained by the assumption of the weak attractive force described above. This force usually involves the lattice vibrations and is between pairs of electrons; these Cooper-pairs have equal and opposite spins and momenta in their centre of mass system. The bound pairs have momenta of the form $(\boldsymbol{p}_i+\boldsymbol{P}/2)\uparrow$, $(-\boldsymbol{p}_i+\boldsymbol{P}/2)\downarrow$ where $\boldsymbol{P}$ is the net momentum of the pair and the arrows refer to spin up and spin down respectively. The range of the attractive force is typically $\sim 10^{-6}$ m, which is greater than the range of the Coulomb repulsive force, $\sim 10^{-8}$ m. The range of the wave-function associated with the pairs of electrons is greater than the average distance between the Cooper-pairs. This overlapping of the pair waves allowed the true many-body approach that was required for a proper theoretical treatment, since the overlapping resulted in the motions of all of the pairs being correlated. It is thus incorrect to think of discreet Cooper-pairs.

The motion of the centre of mass of the Cooper-pairs may be described by a wave-function which has the phase term

$$\psi_p = \exp i(\boldsymbol{P}.\boldsymbol{r}/\hbar - 2\pi\nu t) \tag{1}$$

where $\boldsymbol{P}$ is the net momentum of a Cooper-pair at position $\boldsymbol{r}$ and ν is related to the total energy of the pair by $E = h\nu$. Equation (1) is a travelling wave that represents the motion of the centre of mass of a pair and retains its coherence over indefinitely large distances in a single simply connected superconductor; in the case of a superconducting solenoid the distance may be several kilometres.

The concept of phase in a superconductor is in many ways similar to that of potential in a normal conductor (see page 28). Thus, if there is no applied magnetic field, the phase is everywhere the same where there is no current flow.

Cooper-pairs, being Bose particles, tend to be collectively in the same state, so that although in normal conductivity individual electrons can be scattered out of the regular flow, in a superconductor this scattering is highly improbable. A supercurrent once started therefore tends to continue. The binding energy of a Cooper-pair is small $\sim 10^{-3}$ eV, so that superconductivity occurs at low temperatures where thermal vibration effects are small. Only a relatively few of the more than 900 superconductors listed by Roberts[(29)] have transition temperatures above 5K and, although the intensive search continues, about 20K is the highest transition temperature reported to date. The involvement of the phonon interaction in superconductivity explains why the good normal conductors such as copper are not superconductors, since in good conductors the lattice interaction is weak.

(b) *The Energy Gap*

The energy gap is an important concept in superconductivity. In a normal conductor the electrons may be treated by regarding them as a gas obeying the Fermi-Dirac statistics (the quantum statistics of spin $\frac{1}{2}$ particles). At zero kelvin and in zero electric field, all levels below a certain energy—the Fermi energy E_F—are occupied, and those levels immediately above E_F are empty. As the absolute temperature T is increased, the difference between occupied and unoccupied levels becomes less distinct, occupying a region of roughly kT about E_F (k being the Boltzman constant). In a superconductor there is a forbidden energy zone of width $\Delta(T)$ on either side of E_F in the spectrum of allowed energy values, and a corresponding increase in the density of states immediately on either side of the gap. The energy gap $2\Delta(T)$ may be thought of as being the Cooper-pair binding energy.

(*Note:* This representation is derived from semiconductor theory; there is also the quasi-particle representation, see page 19.)

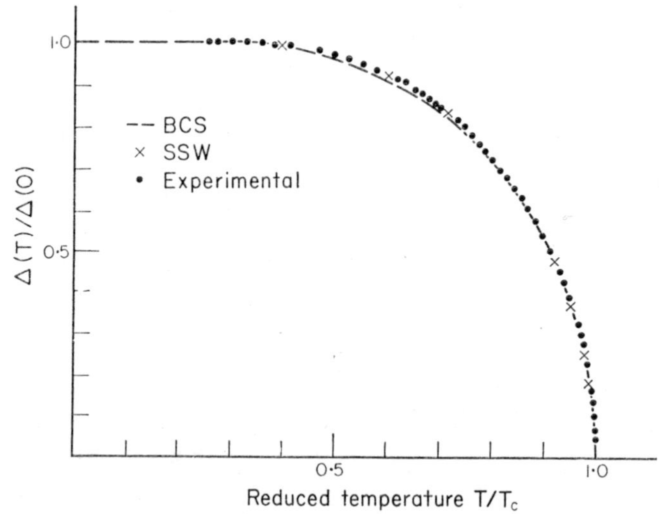

Fig. 2 The variation of the energy gap in lead with temperature, indicating the differences between theory and experiment (after Gasparovic, et al.[30] (*1966*).

The value of $\Delta(T)$ is given to a good approximation on the B.C.S. weak-coupling theory by

$$\Delta(T)/\Delta(0) = \tanh\,[T_c\Delta(T)/T\Delta(0)] \qquad (2)$$

where $\Delta(0)$ is the value of the energy gap at zero kelvin and T_c the transition temperature. The energy gap is thus a maximum at zero kelvin and zero at the transition temperature. According to the B.C.S. theory $2\Delta(0)/kT_c \sim 3{\cdot}5$, and experimentally it is found that for most superconductors $2\Delta(0)/kT_c$ lies within $\pm 0{\cdot}5$ of the predicted value. Equation (2) has been verified experimentally many times and Fig. 2 shows some results obtained by Gasparoviic, *et al.*[30] The small discrepancies between the B.C.S. theory and experiment have been resolved by the strong coupling theory of Swihart, Scalapino and Woda.[31]

(iii) PHASE DIFFERENCE, FLUXOID AND FLUXON

(a) *The Phase Difference Between Two Points*

The pair wave function of equation (1), for a one dimensional wave, has the form (Rose-Innes and Rhoderick,[24] Chapter II):

$$\psi_p = \sin 2\pi(x/\lambda' - \nu t) \tag{3}$$

where λ' the de Broglie wavelength, is related to the momentum $\boldsymbol{P}$ by $\lambda'\boldsymbol{P} = h$ and the frequency ν to the total energy E by $E = h\nu$. In the presence of a magnetic flux $\boldsymbol{B}$ the effects of Lorentz force, $2e\boldsymbol{v} \wedge \boldsymbol{B}$, are automatically included if the expression for the momentum contains an additional term $2e\boldsymbol{A}$, to the product of the pair mass, $2m$, and the velocity, that is if

$$\boldsymbol{P} = 2m\,\boldsymbol{v} + 2e\,A \tag{4}$$

where $\boldsymbol{A}$ is the *magnetic vector potential* defined by

$$B = \text{curl}\,A \tag{5}$$

and e the electronic charge. The quantity $\boldsymbol{P}$ is then conserved in an interaction involving a change of the vector potential $\boldsymbol{A}$ whereas $m\boldsymbol{v}$ alone would not be conserved. An important property of $\boldsymbol{A}$ is that the integral of $\boldsymbol{A}$ around a closed path is, by Stokes' theorem, equal to the total magnetic flux enclosed by that path.

Consider now two points X and Y in a superconductor not necessarily in a straight line path (Fig. 3(a)). If P is zero, i.e. there is no current flowing and zero magnetic flux, the de Broglie wavelength becomes infinite and there is zero phase difference between X and Y; otherwise the phase difference is given by

$$(\Delta\phi)_{X-Y} = \phi_X - \phi_Y = 2\pi \int_X^Y \frac{\hat{\boldsymbol{x}}}{\lambda'} d\boldsymbol{l} \qquad (6)$$

where $\hat{\boldsymbol{x}}$ is a unit vector in the direction of the wave propogation and d is a length element of the path joining X and Y.

The relation of the pair velocity $\boldsymbol{v}$ to the current density J_s is

$$J_s = \tfrac{1}{2} n_s 2e\boldsymbol{v} \qquad (7)$$

where n_s is the density of superconducting electrons, so that the pair density is $\frac{1}{2}n_s$. Hence from equations (3) and (4)

$$\frac{\hat{\boldsymbol{x}}}{\lambda'} = \boldsymbol{P}/h = \left[\frac{2m}{n_s e} \cdot J_s + 2e\boldsymbol{A}\right]/h \qquad (8)$$

and equation (6) becomes

$$(\Delta\phi)_{X-Y} = \frac{2\pi}{h} \int_X^Y \left(\frac{2m}{n_s e} J_s + 2eA\right) d\boldsymbol{l}$$

$$= \frac{4\pi m}{n_s e h} \int_X^Y J_s \,.\, d\boldsymbol{l} + \frac{4\pi e}{h} \int_X^Y A \,.\, d\boldsymbol{l} \qquad (9)$$

The phase difference between X and Y is therefore the sum of a term due to the supercurrent and a term due to the magnetic flux.

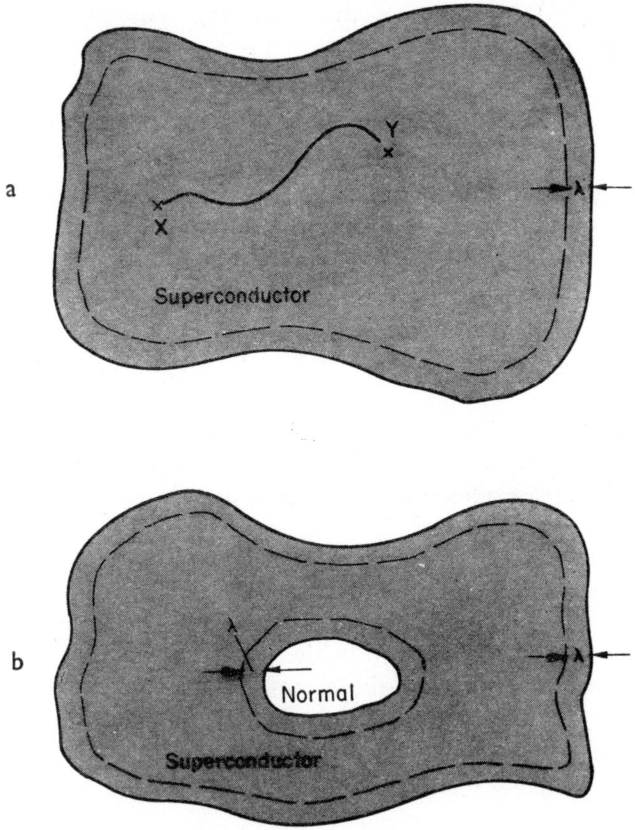

Fig. 3 Illustrating the integration paths used to evaluate the phase difference; (a) *the path between* X *and* Y *and* (b) *the closed path* XY (*not shown*) *lies wholly within the dashed lines and contains a normal region.*

(b) *The Fluxoid*

Consider now the case when the path XY forms a curve which encloses a non-superconducting region, so that X and Y are copunctal, as in Fig. 3(b). The net phase change in going round the loop must be $2n\pi$, where n is an integer, for if this were not so the wave at a particular point would not be single valued. Hence for a closed curve $(\Delta\phi)_{XY}=2n\pi$, so that dividing equation (9) throughout by $h/4\pi e$, yields:

$$\frac{m}{n_s e^2}\oint J_s \cdot dl + \oint A \cdot dl = nh/2e \qquad (10)$$

The quantity on the left hand side was termed the fluxoid by F. and H. London to distinguish it from the flux Φ and was given the symbol Φ'. Hence from equation (10), the fluxoid

$$\Phi' = \frac{m}{n_s e^2}\oint J_s \cdot dl + \int_S\int B \cdot dS \qquad (11)$$

is quantised, where the vector potential term has been replaced by the equivalent magnetic flux term integrated over the area enclosed by the loop X to Y in Fig. 3(b).

(c) *The Quantum of Magnetic Flux or Fluxon*

The first term in equation (11) is generally small, except close to the boundary between a superconducting and a normal region, that is within the penetration depth. An equivalent statement to the quantisation of the fluxoid, therefore, is that in general the magnetic flux enclosed by the curve should only exist in multiples of a quantum, the fluxon Φ_0, given by

$$\Phi_0 = h/2e \simeq 2{\cdot}07 \times 10^{-15}\ \text{Wb}. \qquad (12)$$

The magnitude of Φ_0 is of course very small; for example, a human hair in the earth's magnetic field would enclose several magnetic flux quanta. It is however of considerable importance in the Josephson effects and in describing the properties of type-II superconductors.

The flux quantisation was first observed experimentally by Deaver and Fairbank[32] and at about the same time by Doll and Nabauer.[33] Their experiments demonstrated that the magnetic flux enclosed by a superconducting cylinder increased in units of Φ_0 in striking confirmation of the B.C.S. theory.

C. TUNNELLING OF ELECTRONS THROUGH A BARRIER

(i) ELECTRON TUNNELLING

(a) *Superconductive Electron Tunnelling*

Outside a superconductor, the Cooper-pair wave of equation (3) decays exponentially with distance as exp $(-x/\rho)$, where the length ρ is $\sim$0·1 nm. If, however, the gap between the two superconductors is of the order of 1 nm there is a possibility of an exchange of Cooper-pairs between the two superconductors, and this current results in the Josephson tunnelling effects.

(b) *Quasi-particle Tunnelling*

The "normal" conduction unpaired electrons, or quasi-particles, may also tunnel through the barrier up to rather greater thicknesses $\sim$10 nm, than the Josephson tunnelling supercurrent. This current flows in parallel with the Josephson supercurrent, and hence is always observed experimentally. Figures 4(a) and 4(b) illustrate some quasi-particle tunnelling characteristics that were obtained by Taylor,[36] for similar and dissimilar superconductors on either side of the barrier respectively.

The quasi-particle current is seen from Fig. 4 to be temperature dependent. This current is small when the potential difference is less than the sum of the energy gaps in the two superconductors, but beyond this p.d. the current increases sharply as the Cooper pairing is destroyed. For dissimilar superconductors, a peak also occurs in the quasi-particle tunnelling current when the potential difference is equal to the difference of the two energy gaps, in addition to the sharp increase in current at and beyond the sum of the two energy gaps. The tunnel junction

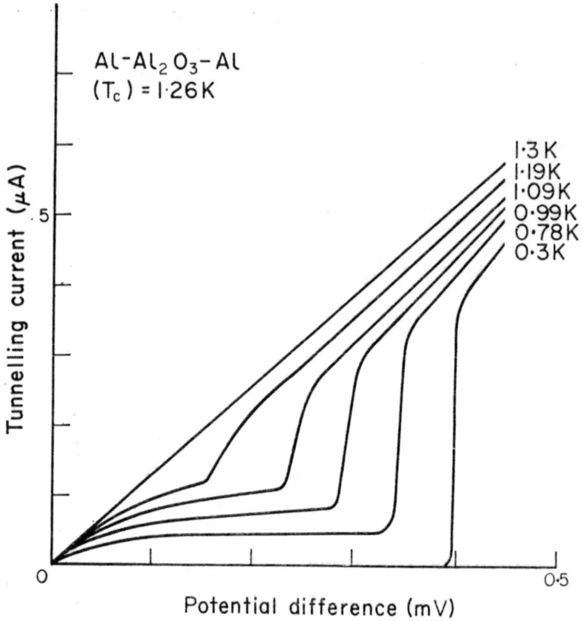

Fig. 4(a) *The quasi-particle tunnelling current between similar superconductors at various temperatures. The sharp rise in current occurs at a p.d. equal to twice the energy gap* (*after Taylor,*[34]).

current–voltage characteristic provides a simple experimental method of obtaining the energy gap of a superconductor. This method of obtaining the energy gap was pioneered by Giaever,[35] *et al.*

(ii) ENERGY LEVEL DIAGRAM

Energy level diagrams are often used to illustrate tunnelling effects qualitatively, and two ways of representing the energy levels are used widely in the literature.[24] These are the quasi-particle representation used in describing effects in superconductivity, discussed by Schreiffer and Wilkins,[36,37] and the band representation (page 13) derived from semiconductor theory.[38] Although the energy diagrams look different they are largely equivalent. In the semiconductor representation the energy bands

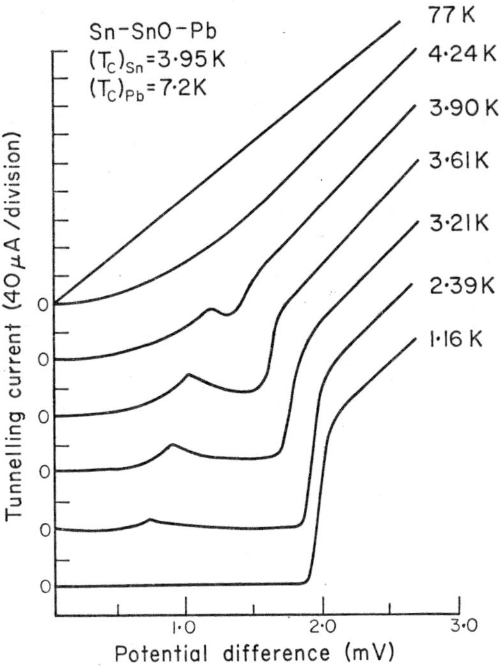

Fig. 4(b) *The quasi-particle tunnelling current between dissimilar superconductors at various temperatures. The additional peak occurs at a p.d. equal to the difference between the respective energy gaps (after Taylor,*[34]*).*

have a meaning different from that of the bands in a normal semiconductor. The allowed tunnelling transitions are made horizontally.

In the quasi-particle representation all electron pairs reside at the Fermi-energy E_F, and the normal fluid comprises the quasi-particles or unpaired electrons. These behave in many respects similarly to electrons in a normal conductor, in that they may be scattered by the lattice etc. However, unlike normal electrons, the energy of electrons in the condensed state is affected by whether or not another energy state is occupied. Thus a quasi-particle of momentum p_i prevents a condensed pair occupying the state $(p_i)\uparrow, (-p_i)\downarrow$. Where reference is made to "holes" these refer to unoccupied energy states and not to charge. Since the quasi-particles behave in many respects like electrons, their energies can be represented by a continuum of levels separated from the level representing the Cooper-pairs by $\Delta(T)$.

Figure 5 illustrates the two types of representation for single electron tunnelling between two superconductors when their potential difference is equal to the sum of their respective energy gaps. The two tunnelling processes shown in Figure 5(a) occur independently.

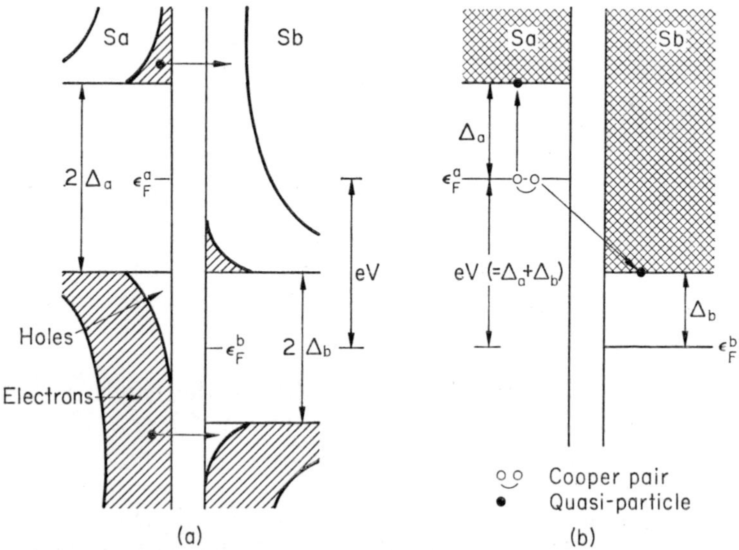

Fig. 5 Diagrams representing the tunnelling of normal electrons between two superconductors when their p.d. is equal to the sum of the energy gaps in the two superconductors; (a) *semiconductor and* (b) *the quasi-particle representation from superconductivity.*

2. The Basic Equations describing the Josephson Effects

A. THE EQUATIONS

Josephson derived a set of essentially five differential equations which related the pair phase and its derivatives to the magnetic fields and voltages in the barrier.

(i) *The Josephson Equations*

Neglecting dissipative processes, these are for a two dimensional barrier (Fig. 6):

$$\partial H_y/\partial x - \partial H_x/\partial y = j_z + C\partial V/\partial t \qquad (13)$$

$$\partial \phi/\partial x = 2\pi d \,.\, B_y/\Phi_0 \qquad (14)$$

$$\partial \phi/\partial y = 2\pi d \,.\, B_x/\Phi_0 \qquad (15)$$

$$\partial \phi/\partial t = (2e/\hbar)V \qquad (16)$$

$$j_z = j_1 \sin \phi \qquad (17)$$

where H_y, B_y etc., are the appropriate components of the magnetic field and flux density, ϕ is the pair phase, C is the capacitance of the barrier per unit area, j the current density, and d the sum of the barrier thickness l and the penetration depths λ in the two superconductors on either side of the barrier.

The detailed derivation of these equations has been given by Josephson and others. However, the first equation is simply the Maxwell equation, curl $H = j_z + \partial D/\partial t$, applied to the barrier conditions, where D is the displacement current term; since the normal component of D between the plates of a capacitor is equal to the charge per unit area, equation (13) follows. The second pair of equations, (14) and (15), result from an application of equation (9) to the closed integration loop (ABCD) shown in Fig. 6 as follows. The difference between the phase change across the barrier at P_1 and P_2 may be written

$$\begin{aligned}\phi_{P_1} - \phi_{P_2} &= [\phi_D - \phi_A] - [\phi_C - \phi_B] \\ &= [\phi_D - \phi_C] - [\phi_A - \phi_B] \qquad (18)\end{aligned}$$

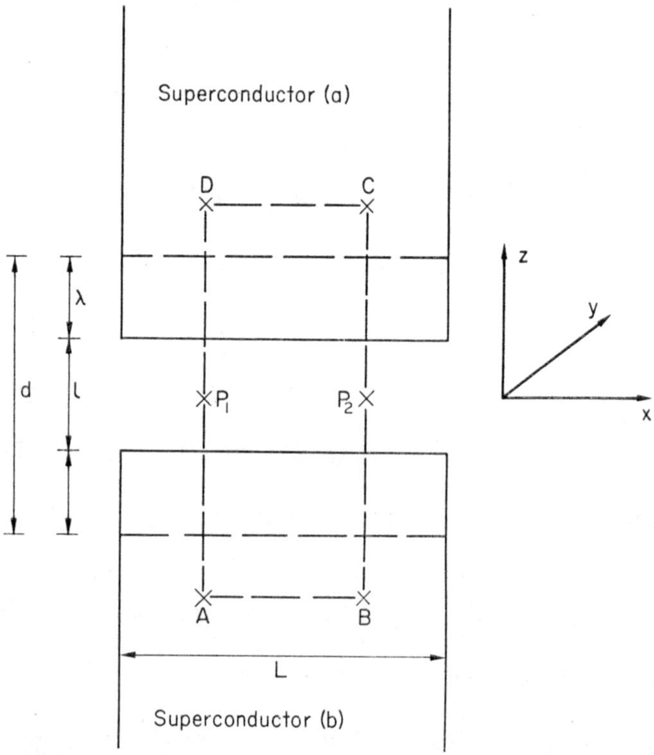

Fig. 6 Illustrating the axes, integration contour and parameters for the two dimensional tunnel junction discussed in the text.

Applying equation (9), the right hand side of equation (18) may then be shown to reduce to $2\pi\Phi/\Phi_0$ thus

$$\phi_{p_1} - \phi_{p_2} = 2\pi\Phi/\Phi_0 \qquad (19)$$

Equations (14) and (15) are then obtained by differentiation of equation (19). The fourth Josephson equation, equation (16), is essentially a result of the quantum relation $E = h\nu$ where the energy $E = 2$ eV in this case. Equation (16) is of great importance in the Josephson effects and is discussed in greater detail in section 2B(ii).

The last equation can only be obtained from the detailed macroscopic theory of the Josephson effects in the barrier, for the magnitude of the supercurrent is determined by the dependence of the barrier free energy on the phase term ϕ. However, if, for example, in the zero voltage case it is assumed that j_z is a function of ϕ so that $j_z = j_z(\phi)$, it is possible to conceive that $j_z(\phi)$ could be a periodic function owing to the angular nature of the phase ϕ

(ii) *The General Barrier Phase Equation*
Using equations (13) to (17) to eliminate the magnetic flux terms leads to the important equation

$$\nabla^2\phi - \frac{1}{v^2}\,\partial^2\phi/\partial t^2 = \lambda_J^{-2}\sin\phi \qquad (20)$$

where $\quad v = 1/(\mu_0 C d)^{\frac{1}{2}},\ \lambda_J^{2} = \hbar/(2\mu_0 j_1 e d)$

and $\quad \nabla^2 = \dfrac{\partial^2}{\partial x^2} + \dfrac{\partial^2}{\partial y^2}$

Taking as typical values $j_1 = 10^6$ A m^{-2}, $d = 10^{-7}$ m, and $C = 2 \times 10^{-2}$ F m^{-2} it is apparent that $v \sim 2 \times 10^7$ m s^{-1} and $\lambda_J \sim \frac{1}{2}$ mm. The Josephson penetration depth λ_J depends on j_1 and this current density in turn depends on the barrier normal state resistance R, see equation (22).

Equation (20) is a general equation for the phase, and it was used by Josephson to investigate many properties of barriers. Since it is a non-linear differential equation, only approximate numerical solutions may be found; however, for small ϕ the linearised version:

$$\nabla^2\phi - \frac{1}{v^2}\,\partial^2\phi/\partial t^2 = \lambda_J^{-2}\phi \tag{21}$$

is obtained, which has exact analytical solutions. The various Josephson effects arise as solutions of these differential equations under different conditions.

The constant j_1 in equations (17) and (20) has the dimensions of a current density and is a measure of the tunnelling probability between the two superconductors (as is readily apparent in Feynman's[20] simplified treatment). It varies exponentially with the thickness of the insulating layer and is also a function of temperature. Ambegaokar and Baratoff[39] have shown that for identical superconductors on either side of the barrier.

$$j_1(T) = (\pi\Delta(T)/2eRA)\ \tanh\,[\Delta(T)/2kT] \tag{22}$$

where A is the area of the junction and R its normal state resistance at low voltages and temperatures. If the two superconductors are dissimilar the expression becomes more complex, and usually the values must be evaluated numerically. For typical junctions $R \sim 10^{-2}\,\Omega$, and $A \sim 5 \times 10^{-8}\,\mathrm{m}^2$, so that j_1 can be as high as $1{\cdot}5 \times 10^6$ A m^{-2}. Riedel[40] has shown that j_1 is also a function of the potential across the barrier. Incidentally, improved fabrication techniques have led to the construction of barriers for particular applications which have lower values for R and hence smaller λ_J and larger j_1 than the above values. Thus Goldman and Kreisman[41] and Schroen and Pritchard[42] have deduced that $\lambda_J \sim 15\ \mu$m with very thin lead–oxide barriers.

B. THE CUT-OFF FREQUENCY AND THE TIME RATE OF CHANGE OF PHASE

(i) THE CUT-OFF FREQUENCY

Substituting the plane wave solution $\phi = \exp i(\omega t - k' . r)$, in equation (20) yields

$$(\omega/\omega_0)^2 = (k' . \lambda_J)^2 + 1 \qquad (23)$$

where

$$\omega_0 = (v/\lambda_J) = (2ej_1/\hbar C)^{\frac{1}{2}}$$

Equation (23) is of a type that occurs frequently in physics, for example in a plasma and also in the waveguide propagation of electromagnetic waves. It follows that, just as in a waveguide, there is a lower limit for ω (namely $\omega = \omega_0$) below which k' does not have real values and hence plane wave solutions do not exist. Using the values given above for j_1 and C, this frequency is $\sim$ 7 GHz.

(ii) THE TIME RATE OF CHANGE OF PHASE

Equation (16) is of considerable importance in the study and application of the Josephson effects, and in this section a little more detail of the background to its derivation is sketched in.

As indicated in Section 1(B), both superconductivity and the superfluid properties of liquid helium II were recognised by London as exhibiting long-range ordering and quantum properties on a macroscopic scale. The long-range ordering is described by an order parameter which persists up to the critical temperature, and it may be defined (Anderson,(12) Khorana(15)) as the mean value of an operator, ψ (r, t), as

$$\psi = \langle\psi(r, t)\rangle = f(r, t) . \exp\{i\phi(r, t)\} \qquad (24)$$

the average being taken over the region defined by the coherence distance.

In a superconductor $\psi(r, t)$ is essentially the anomalous Green's function, $F(r_1, t; r_2, t)$, which was introduced by Gor'kov[43] to describe the pairing of electrons, while for liquid Helium II it is the wave function of the Bose-Einstein condensate. The phase term ϕ describes the current flow and external field effects, and $(f(r, t))^2$ is essentially the number of particles, N, in the system. The phase is canonically conjugate to N, so that $\Delta N \, . \, \Delta\phi \sim 1$.

As with potential, for example, the concept of the phase ϕ for a system as a whole is physically meaningless; however, if two superfluids N_1, N_2, are connected through a barrier, so that $N_1 + N_2 = N$, where otherwise the particles may tunnel through the barrier in both directions, the relative phase ϕ_1–ϕ_2 of the two samples does have meaning. Since ϕ is canonically conjugate to N it may also be regarded as equivalent to the operator $i\partial/\partial N$. Therefore in operator notation $[\phi, H] = i\partial H/\partial N$, and combining this equation with the equation of motion $\dot{\phi} = [\phi, H]/i\hbar$ yields $\dot{\phi} = \hbar^{-1}\partial H/\partial N$, where H is the Hamiltonian of the system. Taking averages, this equation may be written in terms of the chemical potential $\mu(\equiv\partial E/\partial N)$ as

$$\langle \partial\phi/\partial t \rangle = \mu/\hbar \tag{25}$$

If a chemical potential difference $(\mu_1$–$\mu_2)$, is applied between two superfluid samples, the phase difference changes by 2π in a time $h/(\mu_1 - \mu_2)$. If the system has only one state for each ϕ it should return to its original state whenever ϕ increases by $2\pi n$, n an integer. These considerations lead to the equation $\dot{\phi} = (\mu_1 - \mu_2)/\hbar$ or, in the case of two superconductors, to $\dot{\phi} = (2e/\hbar)V$, that is equation (16). This equation was shown to be exact for systems in equilibrium by Anderson *et al.*[44] and correction terms can for most purposes be ignored (see 7(A)).

3. Josephson Effects without a Voltage across the Barrier

A. THE EFFECT OF A MAGNETIC FIELD

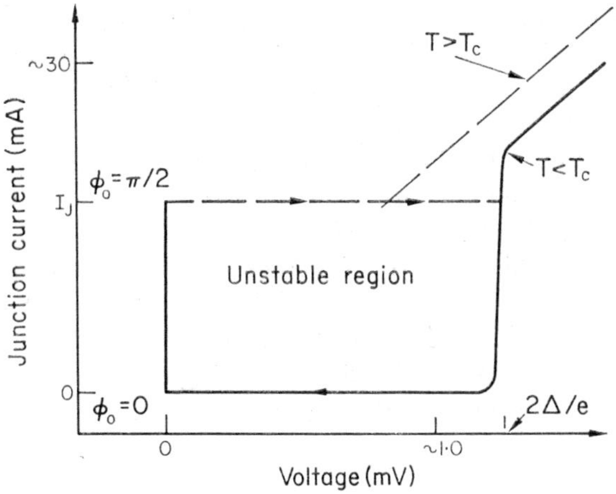

Fig. 7 The simplest Josephson effect. A supercurrent flows across the barrier, the phase difference increasing with current until at a current I_J the tunnelling supercurrent can no longer be sustained. The junction then switches very rapidly to the quasi-particle tunnelling characteristic. There is hysteresis and the portion of the unstable region observed depends on the circuit impedances and noise.

In the absence of a magnetic field or potential difference across the barrier, equation (17) may be integrated over the area of the barrier to give the total current through the junction as

$$I = I_J \sin \phi_0 \tag{26}$$

where ϕ_0 is the phase difference across the barrier. The phase difference across the barrier therefore adjusts until

$I_J \sin \phi_0$ is equal to the total current flowing. If, however, an attempt is made to pass a supercurrent in excess of I_J through the barrier, the phase difference can no longer adjust and a supercurrent can no longer flow. The current–voltage characteristic of the barrier should therefore be as in Fig. 7, which shows typical experimental values.

The existence of the supercurrent through the barrier given by equation (17), although observed experimentally, would not in itself be regarded as confirming the Josephson effects, since a similar current–voltage characteristic could be seen if a thin whisker of supercurrent bridged the barrier and had a low critical current beyond which it became resistive. The behaviour of a barrier showing Josephson effects in the presence of a magnetic field is however quite distinct. The behaviour of the barrier may be divided into the effects of an externally applied magnetic field and the effects resulting from the self-field of the current flowing through the barrier.

(i) THE EFFECT OF AN EXTERNALLY APPLIED MAGNETIC FIELD

As might be expected from the discussion on page 15, there is an alternative form of equation (17) which is frequently used in the literature for the barrier current density when there is a magnetic flux present, namely

$$j = j_1 \sin \phi_0 - \frac{2e}{\hbar} \int_1^2 \mathbf{A} . d\mathbf{l} \qquad (27)$$

where the magnetic vector potential term is integrated across the barrier.

If there is a magnetic flux B_y normal to the plane of the barrier, but no potential difference ($V = 0$), equations (14) and (16) yield

$$\phi = \phi_0 - (2e/\hbar)d \, . \, xB_y \tag{28}$$

Substituting for ϕ in equation (17) gives

$$j = j_1 \sin [\phi_0 - 2\pi(B_y d/\Phi_0)x] \tag{29}$$

It is evident from this equation that the sign of the supercurrent density may change a number of times across the width of the barrier at higher flux densities.

The net supercurrent flowing through the barrier is obtained by integrating equation (29) over the area of the barrier. Thus for a magnetic flux Φ_y through the area of the barrier, the maximum supercurrent that can flow is

$$I_c = I_J \left| \frac{\sin (\pi\Phi_y/\Phi_0)}{\pi\Phi_y/\Phi_0} \right| \tag{30}$$

It was the observation of the type of behaviour of the Josephson current in the presence of magnetic fields indicated by equation (30) by Anderson and Rowell[3] and Rowell[4] that confirmed the existence of the Josephson effects.

In the general case the equation is

$$I = \iint_S j_1 \sin \phi dS$$

$$= \text{Im} \left(\iint_S j_1 \exp \left\{ i \left[\frac{2\pi}{\Phi_0} d(B_y x - B_x y) - \phi_0 \right] \right\} dS \right) \tag{31}$$

and the maximum supercurrent that can flow is represented by

$$I_c = |\textstyle\iint_S j_1 \exp \{i(B_y x - B_x y) 2\pi d/\Phi_0\} \, . \, dS| \tag{32}$$

This expression is the maximum value of equation (31) with respect to changes in ϕ_0, where the integrals are taken over the area of the barrier and Im denotes the imaginary part of the integral.

Josephson[16] pointed out that the integral in equation (31), if regarded as a function of the two dimensional vector (B_x, B_y), is simply the Fourier transform of j_1 with suitable scaling factors and rotated through $\pi/2$. Thus the square of the critical current through the barrier as a function of the magnetic field is similar in shape to the Fraunhofer diffraction pattern observed for electrons, when they are diffracted by an aperture having the same shape as the barrier and transmission coefficient everywhere proportional to j_1.

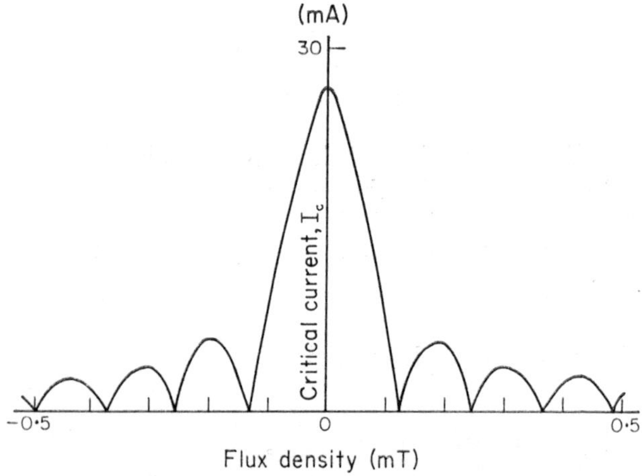

Fig. 8 The variation of the maximum supercurrent I_c as a function of the flux density applied to the barrier; I_c falls to zero each time an integral number of magnetic flux quanta link the barrier.

The variation of I_c with magnetic flux density for a rectangular barrier, as described by equation (30) is shown in Fig. 8. The resemblance to a diffraction pattern is apparent; for typical junctions the periodicity is $\sim 10^{-4}T$ and the maximum value of $I_c \sim 25$ mA. Josephson[16] has pointed out that the analogue with electron diffraction by a single slit is complete, if the diffraction in the latter case is viewed at a fixed point near the slit and the pattern is scanned by varying the magnetic field.

A barrier between two superconductors thus exhibits quantum interference effects. These have been the subject of some elegant experimental work, particularly by Mercereau and his colleagues, and some are considered in later sections of this monograph. The Josephson double junction is of particular interest for it shows interference effects corresponding to electron diffraction by double slits. The effects in multiple and double junctions are discussed in section 5(B).

Since there is a supercurrent flowing through the barrier, it would be expected that, as in a superconductor, there could be a Meissner effect tending to exclude magnetic flux from the barrier, and also a critical field beyond which a supercurrent could no longer flow through the barrier. These and allied effects are discussed in the next section.

(ii) THE MEISSNER TYPE EFFECT OF THE JOSEPHSON SUPERCURRENT

If ϕ is independent of time, equation (20) reduces to

$$\nabla^2\phi = \lambda_J^{-2} \sin \phi \tag{33}$$

and the right hand side of equation (33) may be linearised by putting $\sin \phi = \phi$ assuming that ϕ is small. The resulting equation is similar to London's equation for the Meissner effect in two dimensional form and implies a Meissner effect in the barrier. Thus, in zero or small magnetic fields the Josephson tunnelling currents are confined to a region within $\sim \lambda_J$ of the edge of the barrier and decrease from the edge as $\exp(-x/\lambda_J)$ in the two dimensional case. In this solution of equation (20), λ_J has the character of a penetration depth, which is considerably larger ($\sim \frac{1}{2}$ mm) than the penetration depth in a superconductor, $\sim 10^{-7}$ m. Both Anderson (1963) and Ferrell and

Prange[45] pointed out that this effect would reduce the critical current through the barrier to less than that corresponding to J_1 per unit area, even in zero magnetic field.

As in a superconductor this Meissner effect is destroyed at and above a sufficiently large magnetic flux density B_{c_1} which Ferrell and Prange,[45] Josephson[17] and Anderson[2] showed to have the value

$$B_{c_1} = \Phi_0/2\pi\lambda_J d \tag{34}$$

For typical values this corresponds to flux densities of the order of $10^{-4}T$. Unlike a type II superconductor, as Anderson has pointed out, the passage of a supercurrent across a barrier is not destroyed by fields greater than B_{c_1}, but can flow as long as the superconductors on either side of the barrier retain their phase coherence properties. Thus although B_{c_1} is much smaller than in a superconductor, B_{c_2} in a barrier has a similar magnitude. The flux density B_{c_1} represents the value at which one quantum of magnetic flux moves into the barrier.

(iii) THE SELF-FIELD EFFECT OF THE JUNCTION CURRENT

The current through the barrier itself produces a magnetic field. This self-field contributes to the applied magnetic field, and the magnetic vector potential term in equation (27) is thereby modified. An exact analysis for the effect of the self-field on the behaviour of the barrier is not possible, but an approximate expression for a rectangular barrier has been given by Yamashita and Onodora,[46] Owen and Scalapino[47] and Goldman and Kreisman.[41] The detailed behaviour depends on the junction geometry, but in all cases the self-field effect reduces the total current that flows below current densities corresponding to j_1, by a factor of 0·3 to 0·7.

As examples, if the cross-strip type of junction is used the self-field causes the curves of I_c with flux density to become skew instead of being of the $(\sin\theta)/\theta$ type; Fig. 9 illustrates some results obtained by Yamashita and Onodera.[46] Owen and Scalapino[47] have shown, for example, that with the parallel strip geometry and a strip which is $\sim 10\lambda_J$ wide, Jacobean elliptic functions are involved in the solution. These are of a more triangular shape than the $(\sin\theta)/\theta$ type variation while the successive peaks in I_J are not very well resolved.

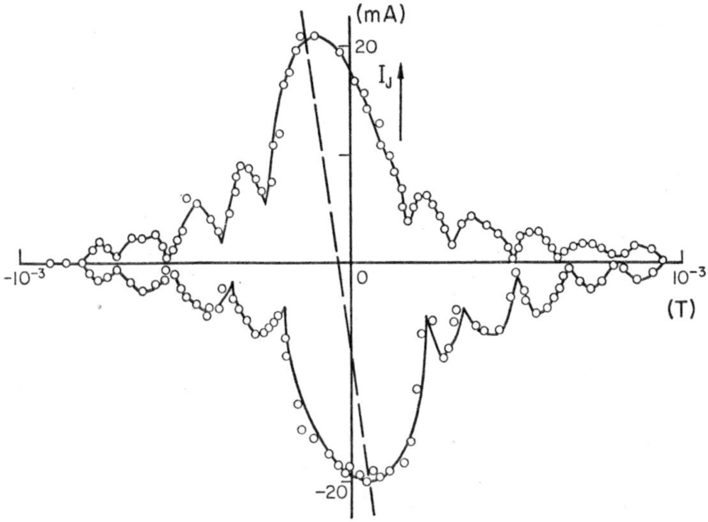

Fig. 9 The effect of the self-magnetic field resulting from the barrier current on the variation of I_J with flux density for a cross-strip tunnel junction (after Yamashita and Onodera,[46] 1967).

The final Josephson effect that occurs with no direct potential difference across the barrier is the plasma resonance that is discussed in the next section.

B. THE PLASMA RESONANCE

The Josephson effects discussed so far have all applied to the situation in which there is no direct voltage across the junction. There is also an effect present under these conditions that involves the alternating tunnelling supercurrent. This effect is the solution of equation (20) (that is $\nabla^2\phi + \frac{1}{v^2}\frac{\partial^2\phi}{\partial t^2} = \lambda_J^{-2} \sin \phi$) for no direct voltage across the junction, that was discussed by Josephson.[18]

If the phase difference makes small oscillations about ϕ_0, so that $\phi = \phi_0 + \delta\phi$, then equation (20) may be approximated by the sum of the Meissner effect equation (33), together with a linearised wave equation in $\delta\phi$, namely

$$\nabla^2(\delta\phi) + \frac{1}{v^2}\frac{\partial^2(\delta\phi)}{\partial t^2} = \frac{\cos \phi_0}{\lambda_J^2} . \delta\phi \qquad (35)$$

The situation for plane-wave solutions is similar to that described in Section B(i), in which the cut-off frequency was discussed ,with the solution satisfying the dispersion relation $\omega^2 = v^2k^2 + \omega_p^2$ where $\omega_p^2 = v^2 \cos \phi_0/\lambda_J^2$ in this case.

The general solution of equation (35) is expected to have solutions corresponding to characteristic resonant frequencies of the barrier. However, there is a solution which does not depend on barrier resonances, namely that at the cut-off frequency $\omega = \omega_p$. This solution has properties[18] which are in many respects similar to those found for ordinary plasma oscillations, for example both the current and electric field are normal to the barrier, while the barrier is a plane of constant phase, so that the disturbance is longitudinal. The plasma oscillations in the barrier involve a periodic interchange of energy between the electric field in the barrier and the tunnelling electrons.

The detection of the plasma resonance radiation is rendered difficult by the severe impedance mismatch between the junction and the microwave cavity, for the impedances differ by a factor of about 2×10^{-5}. However, since the current tunnelling through the barrier is determined by $\sin \phi_0$ and the plasma resonance frequency by $\cos \phi_0$, the plasma frequency depends on the magnitude of the supercurrent and this current may be varied experimentally (at least until $\sin \phi_0 \sim 1$ at $I = I_J$). This property was made use of in the detection of the plasma resonance by Dahm, *et al.*[8] The frequency of the microwave receiver they used to detect the plasma resonance condition was kept constant while the plasma frequency was changed by varying the junction current. They used a narrow band-width receiver in order to improve the sensitivity and signal to noise ratio. Dahm, *et al.*[8] showed that for a small applied magnetic field and small oscillations of the Josephson current, the solutions of the Maxwell equations for the fields within the barrier, together with the Josephson equations, yields an expression for the plasma frequency in terms of the flux through the barrier and the current through it as

$$\omega_p{}^2 = \bar{\omega}_J{}^2 \left| \frac{\sin \pi\Phi/\Phi_0}{\pi\Phi/\Phi_0} \right| \cdot [1 - (I_{\text{d.c.}}/I_{\text{d.c.(max)}})^2]^{\frac{1}{2}} \qquad (36)$$

where $\bar{\omega}_J{}^2 = 2\pi j_1 A/\Phi_0 C$, A being the junction area. If the expression of Ambegaokar and Baratoff[39] for the temperature dependence of j_1 is included, the expression

for $\bar{\omega}_J{}^2$ becomes $\omega_J{}^2(\text{o}) \left[\frac{\Delta(T)}{\Delta(\text{o})} \tanh \left(\frac{\Delta(T)}{2kT}\right)\right]$.

The effects at the barrier are non-linear and Dahm, *et al.*[8] made use of this by applying $\sim 10^{-6}$ W of microwave radiation to it and detecting the second harmonic signal that was generated whenever the plasma resonance condition was traversed. They detected between 10^{-18} and 10^{-17} W of second harmonic power. Their measurement system was capable of detecting as little as 10^{-20} W with the band-width restricted to 0·1 Hz.

Figure 10 shows the results that they obtained for the variation of the plasma frequency with barrier current at different temperatures. The same slope of line was fitted to the higher temperature values as for lower temperatures, since the low temperature results were over a wider range of current. They also observed the dependence of the plasma frequency on the magnetic flux through the barrier over the first two cycles of the diffraction pattern (Fig. 8) and found that the frequency changed as expected from equation (36).

Fig. 10 The variation of the Josephson plasma frequency with barrier current at different temperatures for a tin–tin oxide–tin tunnel junction (after Dahm, et al.,[8] *1968). Note: the same slopes were fitted to the low as high temperature values so that the lines represent a one parameter fit.*

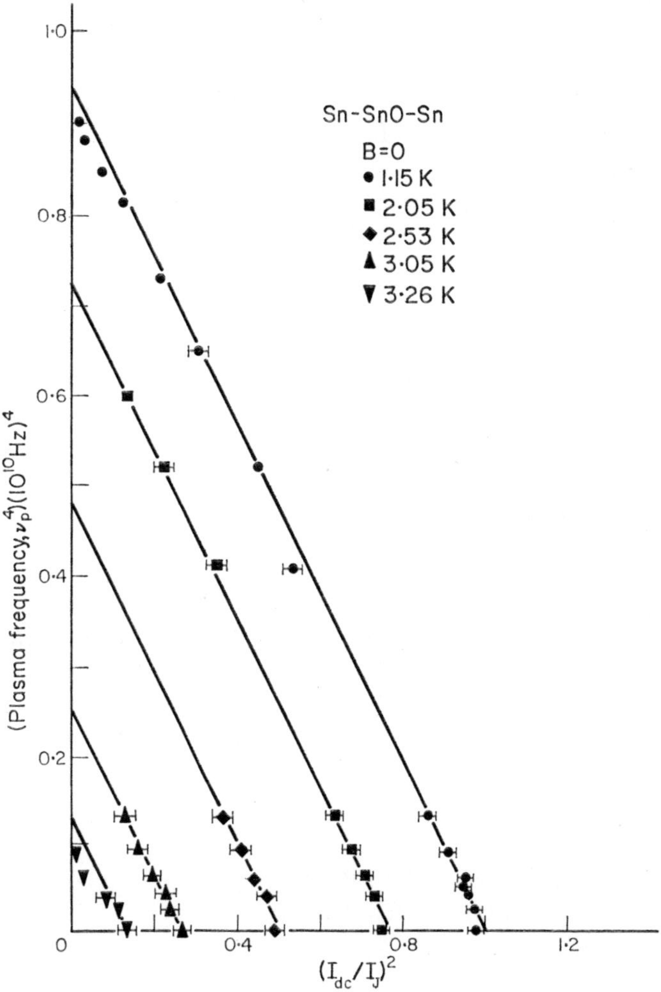
Sn-SnO-Sn
B=0
1·15 K
2·05 K
2·53 K
3·05 K
3·26 K
(Plasma frequency, ν_p^4)(10^{10}Hz)^4
$(I_{dc}/I_J)^2$
1·0
0·8
0·6
0·4
0·2
0
0·4
0·8
1·2

4. Effects with a Direct Voltage across the Barrier

The solutions obtained for equation (20) up to this point have required that the phase difference be either approximately constant, or small. When there is a potential difference across the junction, $\partial\phi/\partial t$ is no longer zero, so that neither approximation is possible. The solutions obtained are therefore rather different and hence the observed effects are different, particularly since energy is now being dissipated in the barrier. However, following Josephson,[16] equation (20) may be written as

$$\nabla^2\phi - \frac{1}{v^2}\frac{\partial^2\phi}{\partial t^2} = \lambda_J^{-2}(j_z/j_1) \qquad (37)$$

together with $j_z = j_1 \sin\phi$

The solutions may then be obtained by an iterative process, that is by first obtaining a solution with the right hand side of equation (37) set equal to zero. This solution is next substituted in the second equation to give an expression for j_z/j_1 which is used in the right hand side of the first equation to obtain the first order solution for ϕ and so on.

A. NO APPLIED ALTERNATING VOLTAGE ACROSS THE BARRIER

(i) VELOCITY MATCHING CONDITION

Equation (16) may be integrated for a constant voltage across the barrier and the solution substituted in equation (17) to give as the zero-order solution

$$j = j_1 \sin (\phi + 2\pi V_0 t/\Phi_0) \tag{38}$$

$$= j_1 \sin \left[\phi_0 + \frac{2ed}{\hbar}(B_y x - B_x y) + \omega_0 t\right] \tag{39}$$

Thus the supercurrent oscillates with angular frequency $\omega_0 = 2eV_0/\hbar$ and, in the presence of a magnetic field, the current distribution varies across the barrier as in a travelling wave which has a velocity equal to $V_0/|B_0|d$. In order to obtain complete solutions of the above equations, they must be combined with the Maxwell equations for the electromagnetic fields within the barrier, since these represent the mechanism for dissipating the energy losses in the barrier. If the effect of the cut-off frequency in equation (23) may be neglected, the phase velocity of the waves is obtained, Langenberg, *et al.*[48], by applying standard transmission line equations to the barrier, as $\bar{c} = 1/(L'C')^{\frac{1}{2}}\varepsilon = c(l/\varepsilon d)^{\frac{1}{2}}$ where ε is the relative permittivity of the barrier dielectric material (e.g. it may be an oxide layer) and L' and C' are the barrier inductance and capacitance per unit length respectively.

Since both a potential difference across the barrier and a current through it must represent a power dissipation in the barrier, it might be expected that when the phase velocity of equation (39) matched that of the electromagnetic waves $\bar{c}$, there could be an interchange of energy between the a.c. Josephson current and the electromagnetic wave in the barrier; i.e., there should be a peak in the d.c. current–voltage characteristic when the condition $V_0/|B_0|d = \bar{c}$ is satisfied. The position of this peak would also vary with the flux density. In fact Langenberg, *et al.*[48] derived a first order expression for the d.c. current density as

$$j = j_1(lj_1/2\varepsilon\varepsilon_0\omega_0 V_0) \cdot 1/Q/(\{1 - [\bar{c}/(V_0/|B_0|d)l]^2\}^2 + 1/Q^2) \tag{40}$$

where Q was the quality factor of the barrier (which had a maximum at the above velocity matching condition). Their results showed that the current maximum occurred at the above velocity matching condition, and that the half width voltage $\delta V \sim \Phi_0 \bar{c}/L$ was $\sim 0{\cdot}2$ mV.

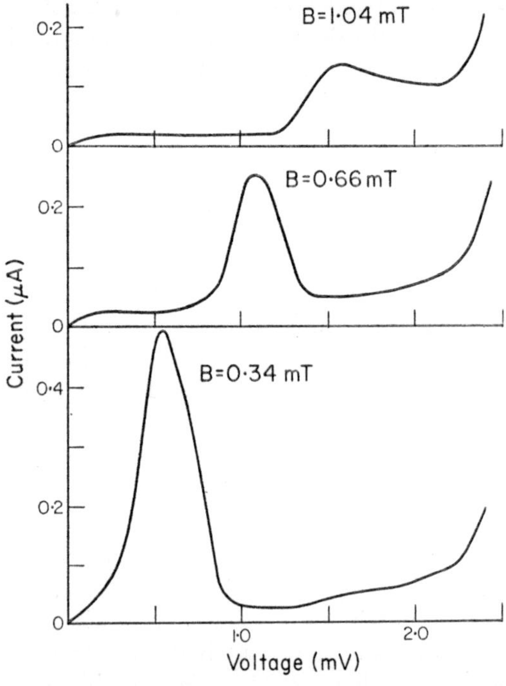

Fig. 11 The peak in the barrier current that occurs at the velocity matching condition, shown for three flux densities (chosen such that the zero voltage supercurrent step was zero), for Pb–Pb *oxide*–Pb *junction at 1·2* K *(after Langenberg,* et al.,[48] *1966).*

The type of resonance curve observed experimentally by Langenberg, *et al.*[48] is shown in Fig. 11 and the corresponding variation of the resonant peak voltage with magnetic field in Fig. 12. It is evident that the approximations made in their simple treatment begin to break down at higher magnetic fields or voltages, where the straight line relationship no longer holds. The experimentally derived values for the Q of the barrier were found to be around 6—rather lower than expected—but, since a cross-strip junction was used, there may have been self-field effects.

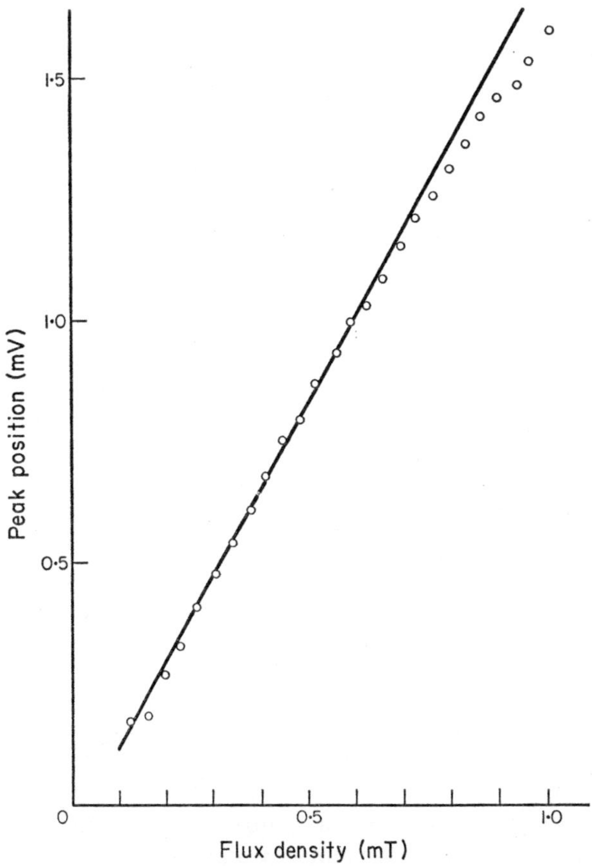

Fig. 12 The variation of the velocity-matching peak voltage with flux density for lead–lead oxide lead junction at 1·2 K. The linear relationship expected from simple theory breaks down at higher flux densities (after Langenberg, et al.,[48] *1966).*

(ii) SELF-INDUCED SUPERCURRENT STEPS

Rather more interesting and important behaviour is observed when the a.c. Josephson current is oscillating at the same frequency as one of the self-resonant frequencies of the barrier. The energy associated with the supercurrent flowing at a finite potential difference across the barrier can couple into the electromagnetic fields in the barrier. If the barrier is of sufficiently uniform construction to have only a few self-resonant frequencies, Langenberg, *et al.*(48) showed that the d.c. current density is given by

$$j_{\mathrm{d.c.}} = \frac{l j_1{}^2}{\varepsilon\varepsilon_0\omega_0 V_0} \sum_{n=0}^{\infty} \frac{1/Q_n}{[1 - (n\pi\bar{c}/\omega_0 L)^2]^{\frac{1}{2}}} \cdot \left[\frac{\sin \pi(\Phi/\Phi_0 \times n/2)}{\pi(\Phi/\Phi_0 - n/2)}\right] \frac{1}{(1 + n\Phi_0/\Phi)^2} \cdot \begin{cases} \frac{1}{2}, & n = 0 \\ 1, & n \neq 0 \end{cases} \tag{41}$$

where $1/Q_n$ applies to the losses in the nth mode. In this case, steps are expected to occur in the barrier current–voltage characteristic, corresponding to the resonant term in equation (41), at equally spaced voltages given by $n(h\bar{c}/4eL)$, where n is an integer. Since the $(\sin \theta/\theta)$ type term in equation (41) depends on the magnetic flux Φ, both the number and size of the supercurrent steps observed in a given I–V plot are determined by the magnetic flux.

These self-induced supercurrent steps are observed by exceeding I_J, then reducing the current, so biasing the junction back along the single electron tunnelling curve. They were first observed by Fiske(6) and are called Fiske-steps. Figure 13 shows a composite plot obtained for several magnetic flux values by Parker, *et al.*(49) Further studies at the University of Pennsylvania have shown that the self-induced current steps may be observed in the apparent absence of any externally applied magnetic field, these zero field singularities decreasing as the temperature was decreased. At the time of writing, their origin is still under investigation. It follows also that if the applied magnetic flux is such that $\Phi/\Phi_0 = n/2$, only those steps associated with n, $n \pm m$, etc., where m is an odd integer, are observed. This behaviour has been confirmed experimentally by Eck[50] (Thesis).

One difficulty in studying tunnel-junctions experimentally arises from that of realising the simple situations envisaged in the theory. The parallel-strip junction of Owen and Scalapino[47] approaches this ideal and has been used in the investigation of the self-induced steps by Chen, *et al.*[51] It seems probable that this type of junction will repay further study, for the self-field effects are more symmetric than in the cross-strip type used for the bulk of the work with tunnel-junctions.

At high magnetic fields many self-resonant modes become possible and irregularities in the junction construction become increasingly important; the steps decrease in height until they merge with the broad resonance of the previous section.

B. DETECTION OF THE A.C. JOSEPHSON SUPERCURRENT

(i) INDIRECT OBSERVATION

Although the phenomena described in the previous section provide strong evidence for the existence of the a.c. supercurrent, it is even more convincing if the current can be observed rather more directly. In the previous section it was stated that when a junction was biased onto one of the self-induced supercurrent steps there was strong coupling, and hence an energy loss mechanism, to the electromagnetic resonances of the barrier. It was expected therefore that it would be possible to observe electromagnetic radiation emitted from a suitably biased barrier. However, the power emitted is small for, as is apparent from Fig. 13, when the junction is biased on one of the

Fiske-steps the current flowing is only a few milliamperes while the direct potential difference is only a few tens of microvolts. The power dissipation in the junction is therefore of the order of a microwatt or less, while several electromagnetic modes may be excited simultaneously.

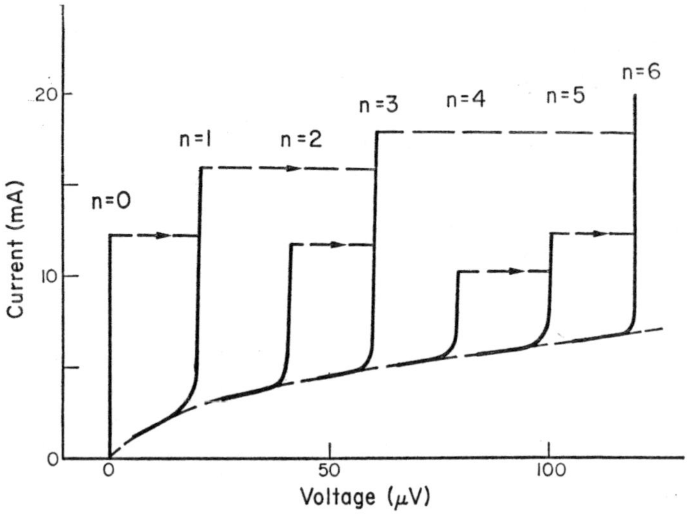

Fig. 13 Composite plot, obtained at different flux densities, of the "Fiske" spikes or steps that appear in the I–V characteristic of a uniform barrier at voltages corresponding to self resonant frequencies of the barrier (via $2eV = h\nu_R$) (after Chen, et al.,[51] *1969).*

Langenberg, *et al.*[48] found that although the junctions were constructed to have self-resonant modes at 9·2 GHz, for the $n = 1$ and $n = 2$ self-induced steps, about five modes were excited simultaneously in practice, thereby reducing the efficiency. This behaviour is qualitatively expected from the non-linear equation (20); unfortunately however, attempts to effect a second order solution giving an indication of their relative amplitudes have so far been unsuccessful. There is also a severe power mismatch between the barrier and any containing waveguide or microwave cavity. (This difficulty was also experienced with the detection of the plasma resonance, discussed in Section 3(B).)

The existence of the emitted radiation was first inferred in a neat indirect experiment by Giaever.[(52)] He formed two Josephson junctions on the same substrate so that their insulating regions were adjacent and the surrounding metal films formed a strip line connecting them. He observed, as would be expected if radiation was being emitted, that when one junction was biased on a self-induced supercurrent step, structure occurred at the appropriate voltages (satisfying $2\,eV = nh\nu$) in the current–voltage characteristic of the other junction. The structure in the characteristic of the second junction is a result of the induced-step effect that is described in Section 4(C). One junction was therefore an emitter and the other a detector of radiation—suggesting two applications of the a.c. Josephson effect, see Section 7.

(ii) DIRECT OBSERVATION OF EMITTED JOSEPHSON RADIATION

The first direct observations of the emitted electromagnetic radiation were made by Yanson, *et al.*[(7)] who detected 10^{-14} W of emitted radiation from a tunnel–junction. Langenberg, *et al.*[(53)] detected up to 10^{-11} W also with tunnel–junctions. A point-contact junction allows better coupling to the microwave cavity and Dayem and Grimes[(54)] obtained up to 10^{-10} W of radiation corresponding to a d.c. to microwave conversion efficiency of $\sim$0·1 per cent. These experiments were all performed in the microwave region at frequencies of about 9·2 GHz. It is rather easier to couple the radiation into a barrier at lower frequencies, for it behaves like a transmission line and the impedance mismatch problem is thereby overcome to some extent. Zimmerman, Cowen and Silver[(55)] obtained a conversion efficiency of $\sim$5 per cent at the much lower frequency of 30 MHz.

The tunnel–junction has self-resonant frequencies depending on the geometry, which are in the microwave region $\sim$10 GHz; it is not, therefore, a broad-band device. The point-contact, on the other hand, has much higher self-resonant frequencies $\sim 10^{12}$ Hz; when it is used at lower frequencies, both the frequency and frequency stability of the emitted radiation are determined by the resonant frequency and Q of the enclosing microwave cavity respectively. The frequency stability of the emitted radiation from a tunnel–junction is dependent on the stability of the current source. Josephson[16] envisaged that the emitted radiation, though coherent, would not necessarily be monochromatic, since the linewidth is affected by Johnson noise from any series resistor. He used a simple equivalent circuit for the junction and suggested that the spectral width would vary as $(1/R_s)^{\frac{1}{2}}$, where R_s is the series resistance.

Langenberg, *et al.*,[48] working at 9·2 GHz with a tunnel–junction, observed a linewidth of 600 kHz with a series resistor of 1 Ω. The linewidth was smaller with a series resistor of 100 Ω, being limited by the 100 kHz frequency modulation of the local oscillator used in the detection system. Later work by the Pennsylvania group, Dahm, *et al.*,[58] considered the contribution of the quasiparticle current I_{qp} to the linewidth of the emitted radiation $\Delta\nu$ and predicted the relation

$$\Delta\nu = (4\pi R_D{}^2 kT/V_0)(2e/h)^2(I_{qp} + I_p) \qquad (42)$$

where R_D was the dynamic resistance of the barrier which was inferred from the slope of the first Fiske step, V_0 is the potential difference, and I_p the pair current. Measurements on Pb–I–Pb, and Sn–I–Sn junctions, covering a temperature range 1·9 to 4·2 K and current range 3 to 150 μA, gave good qualitative agreement with the above expression, although the experimental linewidths were about twice those expected. The linewidth has also been studied by Silver, *et al.*[59]

Since the linewidth of the emitted radiation also depends on the absolute temperature of the series resistor, it suggests an application as a voltage to frequency conversion element in an absolute noise thermometer; this is being actively pursued by several experimentors, e.g. Kamper, *et al.* (60) Some "non-Josephson" radiation from Josephson tunnel–junctions has been reported by Langenberg, *et al.*(61) at frequencies not satisfying $2\ \mathrm{eV} = nh\nu$.

C. RADIATION INCIDENT ON THE BARRIER

Although this appears here as the last effect, the case where there is radiation incident on the barrier has been the most studied experimentally, and is perhaps the most useful a.c. effect at this stage of development. If the voltage across the barrier can be represented by the sum of a direct and an alternating voltage, of the form $V_0 + v_0 \sin \omega t$, equation (16) may be integrated and substituted in equation (17) to give an expression for the current density, j, as

$$j = j_1 \sin\left[\phi_0 + \omega_0 t + (\omega_0 v_0 / V_0 \omega) \sin(\omega t + \theta)\right] \qquad (43)$$

where θ is the phase difference between the Josephson a.c. supercurrent and the applied microwave radiation.

This equation represents an alternating supercurrent that is frequency-modulated by the incident microwaves and has sidebands at frequencies $(2e/h)V_0 \pm n\nu)$, where $\nu = \omega/2\pi$ and n is an integer. If for some value of n the relation $(2e/h)V_0 = n\nu$, is satisfied exactly, there is also a net direct component of the supercurrent (corresponding to the sideband at zero frequency) which is given by

$$j_{\mathrm{d.c.}} = j_1(-1)^n J_n(nv_0/V_0) \sin(\phi_0 - n\theta) \qquad (44)$$

where J_n ( ) represents the nth-order Bessel function. The sign of the direct supercurrent is positive or negative according to the relationship between the pair phase ϕ and the microwave phase θ and corresponds to n-induced photons emitted or absorbed respectively. In effect, a direct supercurrent may tunnel across the barrier whenever the beat frequency between the pair waves on either side of the barrier is harmonically related to that of the incident radiation, the beat frequency being determined by their potential difference via the Plank equation $E = h\nu$. It was the observation of these microwave induced

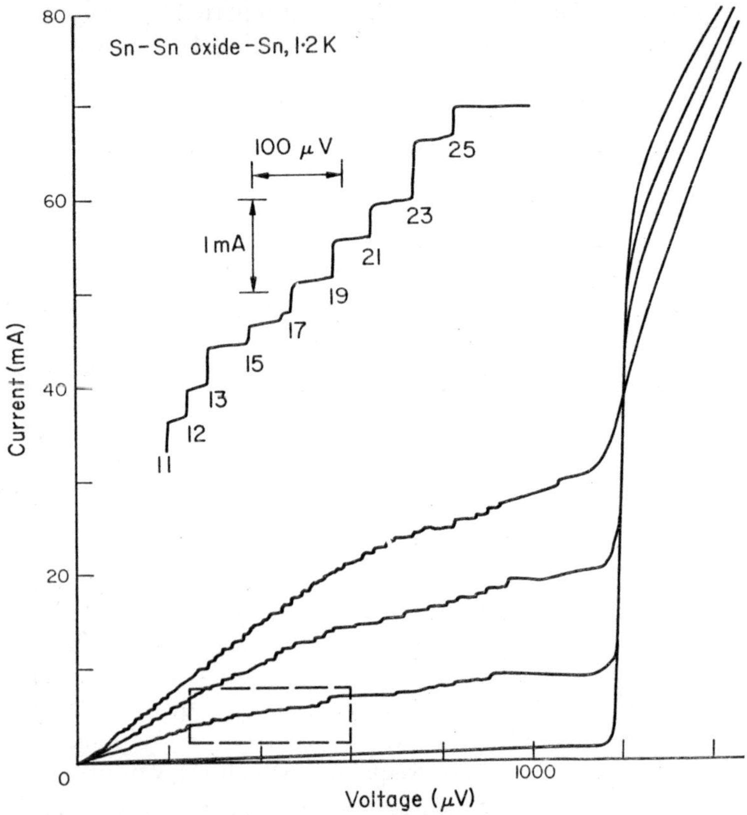

Fig. 14 The supercurrent steps that appear in the barrier I–V characteristic when radiation is incident on the barrier at voltages governed by the quantum relation $2\,eV = nh\nu$. The applied flux and power are such that for the most part only the odd supercurrent steps are visible. As the power is increased the steps move upwards until the curve appears resistive with no visible steps (the latter is not predicted by simple theory) (after Taylor, et al.,[63] *1968).*

supercurrent steps by Shapiro[5,62] that confirmed the existence of the a.c. Josephson current.

The direct current–voltage characteristic of a junction resembles a staircase, with supercurrent steps appearing at constant voltage intervals appropriate to the applied microwave frequency. The current amplitude of a supercurrent step depends on the pair phase difference ϕ_0 and so may be "tuned" by varying the magnetic flux. The step amplitude also depends on the microwave voltage in the barrier and so varies with the applied microwave power.

Figure 14 shows a current–voltage characteristic of a tunnel junction which was obtained by the Pennsylvania group.[63] As is usually the case, the amplitude of the zero-voltage supercurrent step has been reduced to zero by applying an appropriate magnetic field. The slope of the characteristic between the supercurrent steps is affected by the normal conduction or quasi-particle current; the slope resistance may vary in experimental junctions from near infinite for pure tunnel junctions with weak-coupling, to a few ohms for a junction having a shorting bridge of metal.

It is of course very much easier to couple energy into the barrier when the microwave frequency is close to one of the barrier resonant frequencies and these are determined by the dimensions of the barrier. If the phase ϕ_0 varies over the barrier, for example, as a result of the self-field effect or an applied magnetic field, the total current flowing, being the integral of equation (42) over the barrier, becomes quite complicated; see, for example, Anderson[2] and Scott.[64]

If the power is increased sufficiently, the above simple analysis breaks down. Experimentally, it is observed that the characteristic becomes progressively more resistive in appearance until the supercurrent steps eventually disappear.

5. Other Configurations Exhibiting Josephson Effects

A. SINGLE JUNCTIONS

Fortunately perhaps, it is not at all essential to have a purely insulating barrier in order to observe the Josephson effects. It has been seen however that even with an insulating barrier the solutions to the differential equations are only approximate. The difficulties of obtaining solutions are even more pronounced when the barrier is not insulating. Increasing theoretical attention is however being paid to this situation, with quite promising results.

When the barrier is non-insulating—be it semiconducting normally conducting or weakly superconducting—there is a quasi-particle current flowing in parallel with the supercurrent. Following, for example, Waldram, Pippard, and Clarke,[65] equation (17) becomes

$$j = j_1 \sin \phi + \sigma V \tag{45}$$

where σ is the ordinary normal conductance term per unit area. Although in an insulating barrier the impedances are either near-infinite or zero, that is there is only a supercurrent flowing, when there is a conduction current this is no longer true, hence the observed behaviour depends rather markedly on the source impedances. Thus, Clarke, *et al.*[65] show that for a voltage source applied to the junction of the form $V_0 + v_0 \sin \omega t$

$$J = J_1 \sin [\phi_0 + \omega_0 t - (\omega_1/\omega) \cos \omega t] + \sigma V \tag{46}$$

where $\omega_0 = 2eV_0/\hbar$ and $\omega_1 = 2ev_0/\hbar$. If, as is closer to the experimental situation, the junction is supplied from a current source $J = \bar{J}_0 + \bar{J}_1 \sin \omega t$ then

$$J = J_1 \sin \phi + (\sigma\hbar/2e)\,\partial\phi/\partial t$$

hence

$$(\sigma\hbar/2e)\,\partial\phi/\partial t = \bar{J}_0 + \bar{J}_1 \sin \omega t - J_1 \sin \phi \qquad (47)$$

Equation (47) gives very different results from equation (46). The latter current–voltage characteristic. Clarke, *et al.*[65] solved equation (47) numerically and found that there is complex harmonic and sub-harmonic structure in the junction *I–V* characteristic, as is observed experimentally. The sub-harmonic structure depends on the impedance of the a.c. source, disappearing only when it is identically zero.

McCumber[66] has considered the effects of a reactive source impedance on the behaviour of a barrier, and shown that the characteristic differs for an inductive or capacitative source impedance. This change has experimental consequences, for, apart from the impedance of the barrier there is also the impedance of the leads to the junction to consider. Thus, unless the source impedances are identical, the same *I–V* characteristics are not observed with an oscilloscope as with a pen-recorder. They are of course closely similar but, for example, the microwave power for optimum amplitude for a particular step may differ for the two cases.

Werthamer and Shapiro[67] have made an interesting analogue computer study of the behaviour of a barrier when energised by either a constant voltage or constant current source. Their analysis enabled the effects of some of the approximations that are normally made in the theoretical treatments to be studied. The source impedance can also affect the behaviour of an insulating barrier (see Albegova, *et al.*[68])

In discussing the Meissner effect in an insulating barrier, it was mentioned that there is a transition at sufficiently high magnetic flux densities $\sim 10^{-4} T$, to behaviour analogous to that in a type-II superconductor. A valuable concept in type II superconductivity is that of the supercurrent vortex. This vortex represents a circulating supercurrent sufficient to contain one quantum of magnetic flux. The a.c. Josephson current may also be represented by a series of supercurrent vortices traversing the barrier at the appropriate frequency and coupling with external electromagnetic and magnetic fields.

This concept was introduced by Parks and Mochel(69) and Anderson and Dayem(70) to explain the behaviour of the evaporated film weak-link type of junction in magnetic and electromagnetic fields respectively, and has been used by a considerable number of authors. It is particularly useful for comparing the superconducting Josephson effects with those occurring in superfluid helium which are discussed in Section 6.

B. MULTIPLE JUNCTIONS INCLUDING DOUBLE JUNCTIONS

(i) MULTIPLE JUNCTIONS

In Section 3(A) it was indicated that the variation of the critical current of a junction with magnetic field strongly resembled the diffraction pattern of electrons by a single slit. It is natural, therefore, to expect that the multiple Josephson Junction should show effects that are very similar to the diffraction of electrons by multiple slits.

An experiment with multiple junctions was successfully carried out by Zimmerman and Silver,(71) who used up to six point-contact junctions in parallel and observed a much sharper pattern as expected. Clark(72) has used an array of lead spheres to form multiple point-contacts, for investigating infra-red applications of the Josephson effects.

Most of the experimental work and applications of multiple junctions has concerned the double junction system, beginning with the work of Jaklevic, Lambe, Silver and Mercereau.(82) Point-contact Josephson junctions are usually employed, mostly for their smaller area, but also because they have the advantage that they are adjustable, thereby allowing greater success with forming the double junction.

(ii) DOUBLE JUNCTIONS

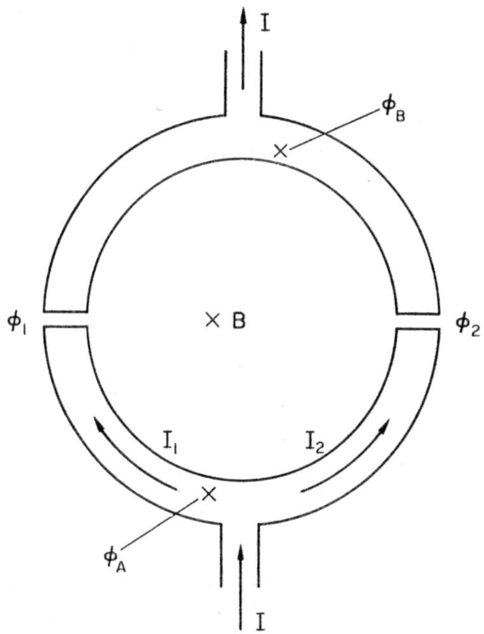

Fig. 15 An illustration of the double junction considered in the text.

If there are two junctions in parallel (Fig. 15) then

$$I = I_1 + I_2 = I_0 (\sin \phi_1 + \sin \phi_2) \qquad (48)$$

where ϕ_1 and ϕ_2 are the phase differences across the two junctions and the junctions are assumed to have the same critical currents and to be of small area. In order that the wave function shall be single valued these phase differences must also satisfy equation (9) applied to the loop containing the two junctions, that is

$$\phi_1 - \phi_2 + \oint \frac{2e}{\hbar} A \, . \, dl = 2\pi n, \; n = 0, \pm 1, \pm 2, \qquad (49)$$

Hence, taking $n = 0$, if Φ is the magnetic flux enclosed by the loop, equation (48) rearranged gives (Zimmerman and Silver[71])

$$I = 2I_0 \cos\left(\pi\Phi/\Phi_0\right) . \sin\left(\phi_A - \phi_B\right) \qquad (50)$$

where ϕ_A and ϕ_B are the relative phases at two arbitrary points on opposite sides of the ring.

The maximum values of the current, corresponding to $\phi_A - \phi_B = \pm\pi/2$, are

$$I_c = 2I_0 \left| \cos\left(\pi\Phi/\Phi_0\right) \right| \qquad (51)$$

From this equation, the peak-to-peak amplitude should be $2I_0$; however, more detailed considerations show that it should be $2I_0$ *or* $\leqq \Phi_0/\mathscr{L}$ whichever is the smaller, where $\mathscr{L}$ is the inductance of the ring.

The occurrence of the latter expression may be understood by noting that it corresponds to the circulating current required to maintain one flux quantum within the area of the ring, a result that would be expected if there were no junctions present and the ring was entirely superconducting. Indeed, the two situations are closely similar, for Goldman, Kreisman and Scalapino[74] and Smith[75] demonstrated that a stable supercurrent could be made to circulate around a ring containing two such junctions when there were one, two or three magnetic flux quanta linking the area of the ring. Persistent currents were observed for periods of up to an hour, a property that provided further strong evidence that the currents through the junctions were superconducting.

(a) *Two Interesting Applications*

(i) *Magnetometer.* The modulation depth of the critical current decreases as the size of the ring increases, as a result of the inductance effect, but at the same time the sensitivity to magnetic flux density increases. Figure 16 illustrates the variation of I_J with flux density that was observed for a double tunnel junction of similar construction to that illustrated in Fig. 20. Larger areas, giving greater sensitivity to flux changes, may be used with point-contact junctions; thus Zimmerman and Silver (1966) used a ring diameter of 1 cm in their experiments and obtained a modulation change of one cycle for a flux density change of $\sim 2 \times 10^{-11} T$.

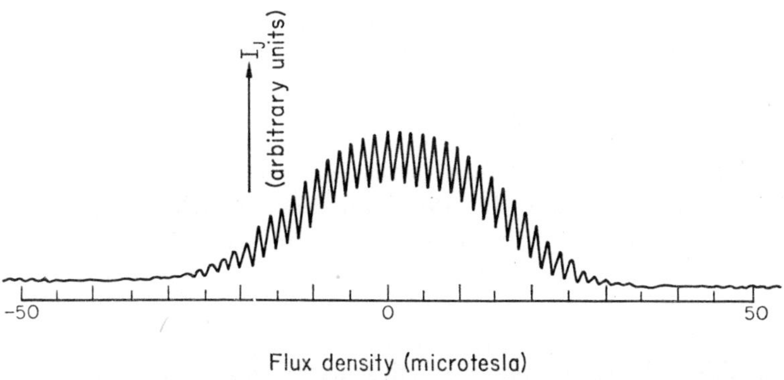

Fig. 16 The variation of the barrier current I_J (the current at which a p.d. first appears across the barrier) as a function of magnetic flux density for a loop containing two tunnel junctions. The broad pattern is the modulation from a single barrier and the fine one the interference between the supercurrents through the two barriers. The suitability as a magnetometer is readily apparent; compare also the flux scale with that of Figs. 8 and 9. The modulation is no longer complete owing to the inductance of the loop (after Jaklevic, et al.,[82] *1965).*

The double Josephson junction can therefore be used as the basis of a very sensitive magnetometer. For example a double-junction has been used by Cohen,[76] *et al.* to detect the magnetic flux associated with human heart-beats, which give $\sim 3 \times 10^{-11} T$ peak signal. The cryostat must be very well screened magnetically and electrically, and the leads to the junction very carefully filtered to use the magnetometer at its full sensitivity.

(ii) *Galvanometer.* The solder-drop junction of Clarke[77] was devised originally for use as a very sensitive galvanometer, and the following are its operating principles. The galvanometer current passes through the Niobium wire (Fig. 17) and produces a magnetic flux through the area between the junctions which modulates the Josephson current. In Clarke's junctions, a current I_g of about 200 μA produced one magnetic flux quantum through the area between the junctions, and a change in this current of as little as $\frac{1}{2}$ per cent could be seen. Therefore, with a resistance of $5 \times 10^{-8}\ \Omega$ in series with the niobium wire, a potential difference of 5×10^{-14} V could be detected at the input terminals.

The time constant of the device, determined by the series resistance R_s and by the inductance of the leads, is typically about a second for the sensitivity quoted above. The sensitivity obtained by Clarke was therefore better than that of about 10^{-12} V achieved by other types of galvanometer by Pippard and Pullan[78] or Templeton,[79] although Ries and Satterthwaite[80] have approached very close to Clarke's sensitivity with the modulated transformer method. The latter method does not use the Josephson effect; it is now available commercially, sensitivity $\sim 4 \times 10^{-14}$ V, as is the SQUID below.

The double junction has been termed a SQUID—*s*uperconducting *q*uantum *u*ndulating *i*nterference *d*evice—by the Zimmerman, Jeklevic, Lambe and Silver group in the magnetometer application. Clarke replied with his

SLUG—*s*uperconducting, *l*ow inductance, *u*ndulating *g*alvanometer!

Two particularly interesting experiments performed with double junctions are discussed in the next section.

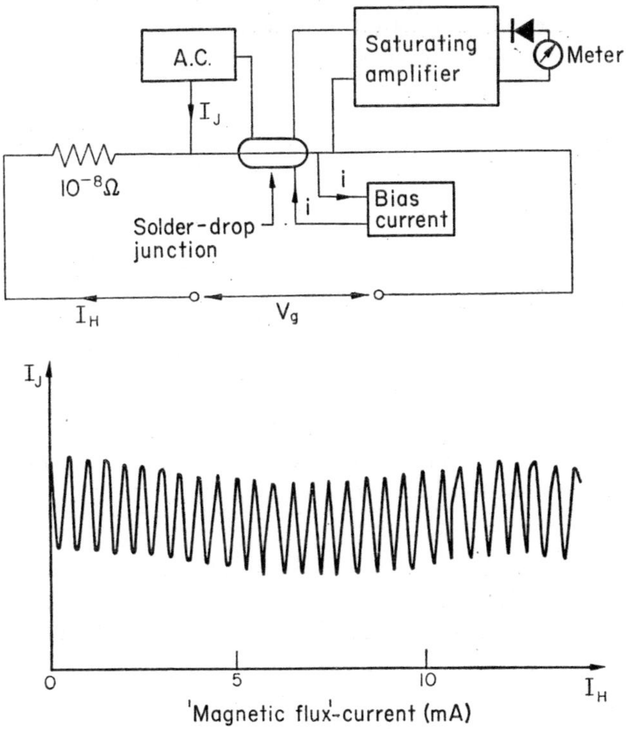

Fig. 17 The solder-drop junction used as a galvanometer, sensitivity 10^{-14} V. The amplifier saturates whenever a voltage appears across the barrier, hence its output increases as I_J decreases (after Clarke, 1966).

(b) *Two Interesting Experiments*

(i) *Reality of the Vector Potential.* One of the interesting debates in quantum mechanics has been whether the magnetic vector potential exists where there is no magnetic field. In order to investigate this, Jaklevic, *et al.*(81,82) constructed a double Josephson junction with a very small solenoid between them, as illustrated in Fig. 18, so that the only significant amount of magnetic field was that within the solenoid and a negligible amount was on the superconducting wires themselves. They found that the junction critical current varied periodically with the field inside the solenoid, even though that field did not touch the junctions, demonstrating the reality of the vector potential.

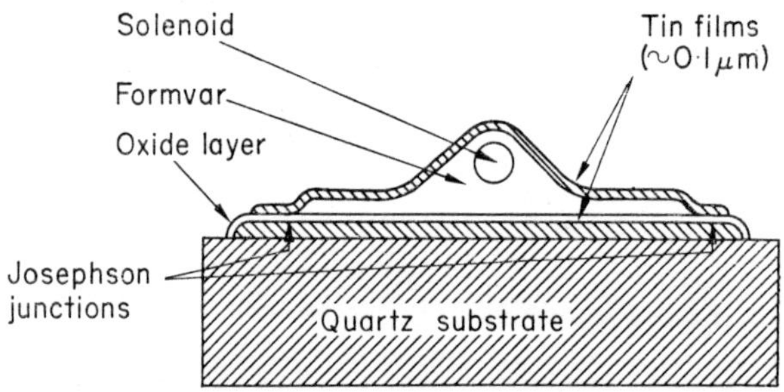

Fig. 18 The tin–tin oxide tin double tunnel junction with embedded solenoid used to investigate the reality of the vector potential **A** *(after Jaklevic,* et al., *1964, 1965).*

(ii) *Compton Wavelength of Superconducting Electrons.* In discussing the phase difference between two points in Section 1, it was found that the difference depended on the electron momentum and a vector potential term. Usually, the mechanical momentum term reduces to one containing the current density, but if there is angular momentum involved this is no longer the case. Zimmerman and Mercereau[83] showed that for a rotating double junction, equation (51) should be replaced by

$$I_{\max} = I_0 \left| \cos\left(\frac{1}{2\hbar} \oint P \, . \, dl\right) \right| \tag{52}$$

where $p = 2(m\boldsymbol{v} + e\boldsymbol{A})$
or, on evaluating the line integral,

$$I_{\max} = I_0 \left| \cos 2\pi \left(\frac{p\omega}{\hbar} + \frac{e}{h} \Phi \right) \right| \tag{53}$$

where p is the angular momentum. If the area of the loop is S and the inteferometer is rotated with angular velocity Ω, then

$$I_{max} = I_0 | \cos 2\pi \{ (2m/h) S\Omega + (e/h)\Phi \} | \tag{54}$$

Zimmerman and Mercereau[(83)] constructed a vanadium double-junction with an area S of $7{\cdot}4 \times 10^{-6}$ m^2, which they rotated at angular velocities of up to about 10 rad s^{-1}. They measured the maximum supercurrent as a function of the rotation rate and deduced that

$$h/mc = (7{\cdot}3 \pm 0{\cdot}3) \times 10^{-4} \text{ J sec kg}^{-1}$$

or that the Compton wavelength h/mc of the electron was

$$h/mc = (2{\cdot}4 \pm 0{\cdot}1) \times 10^{-12} \text{ m.}$$

Their value was in good agreement with the accepted value of about $(2{\cdot}426 \ldots) \times 10^{-12}$ m. The experiment provides additional evidence that there are macroscopic quantum effects in superconductors.

Josephson has pointed out that the mass of the electron in the above expressions should be relativistically corrected by the ratio of the Fermi energy to the rest mass of the free electron. This correction however is not important with the present experimental sensitivities.

C. METHODS OF CONSTRUCTING JOSEPHSON JUNCTIONS

The tunnel-junction has been the most studied theoretically, but Josephson effects may be observed in other configurations where the two superconductors are no longer separated by an insulator. They may be separated by a semiconductor, normal conductor or a thin bridge of superconductor. This section discusses some of the types of junction that that have been evolved, although some of them have already been mentioned. The basic types of single junction are illustrated in Fig. 19.

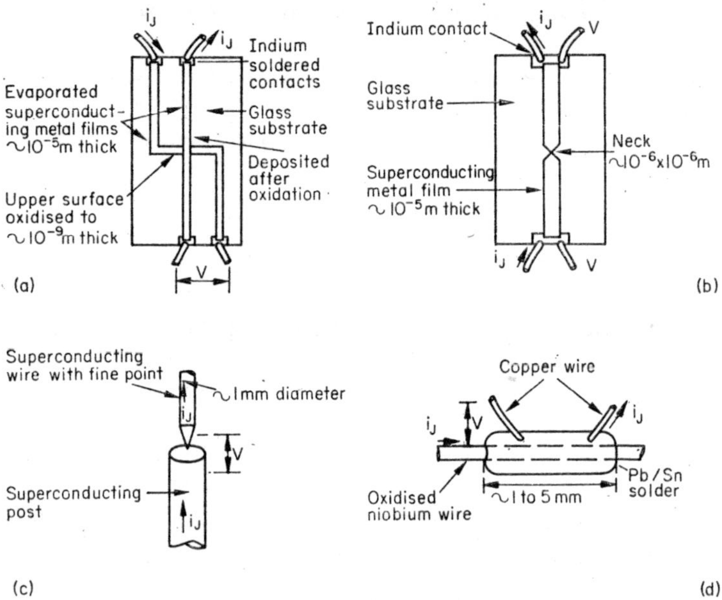

Fig. 19 The principal types of single junction used in experiments on the Josephson effects; (a) *the cross strip tunnel junction (e.g. Pennsylvania group)*; (b) *the thin film weak-link of Dayem and Grimes (1966)*; (c) *the point contact, Zimmerman and Silver*; (d) *the solder drop junction of Clarke (either double or single).*

(i) BARRIER JUNCTIONS

(a) *The Tunnel-Junction*

The tunnel-junction is formed on a substrate; a microscope cover glass is commonly used, but materials such as Rutile have also been employed (Zimmer).[84] The substrate is very carefully cleaned to give good adhesion for the metal films that are subsequently evaporated onto it. The first evaporation is usually that of the four terminations (cf. Fig. 19) for making the external connections; indium, with its low melting point and good adhesion to glass, is often used. (The junctions are normally made as four-terminal devices, that is there are two current and two potential connections.)

Next, a thin strip of superconductor—commonly lead, tin or niobium is evaporated. The evaporated film is then oxidised. The oxidation is preferably carried out in a dry atmosphere, which may be room temperature air or pure oxygen; more rapid oxidation may be achieved by using higher temperatures.

When the oxide thickness is about 10^{-9} m, a final cross-strip of the same or a different superconductor is evaporated. The tunnel-junction is formed where the strips overlap and each strip makes contact, at either end, to one of the pre-evaporated indium strips. A final protective coating of photo-resist is often applied, and junctions may then be cycled to room temperature or kept for long periods at liquid nitrogen temperature.

The junction behaves as a perfect four-terminal device only when it is superconducting, and the room temperature resistance is of the order of 0·1 Ω or may sometimes even appear to be negative, depending on the character of the current distribution where the strips cross one another. The junction normal resistance at low temperature ranges from 10^{-3} Ω to about 1 Ω, depending on the dimensions.

The parallel-strip junction is gaining favour, since the self-field effects are more symmetric, and this type of junction will probably be used increasingly. Parallel-strip junctions, are similar in construction to that of Matisoo[(85)] (Fig. 27) without the control strip.

(b) *Superconductor-Normal Metal-Superconductor Sandwich*
Clarke[98] has used this type of junction in connection with his work on the verification of equation (16) (see Section 7(A)). He used a glass substrate, and evaporated successively a strip of superconductor (lead, tin or indium), a disc of normal metal (usually copper) and finally a cross-strip over the disc at right angles to the first superconductor. The superconductors were about 0·2 mm wide and $\sim 5 \times 10^{-7}$ m thick and the copper was typically 10^{-6} m thick.

(c) *Semiconducting Barrier*
Tunnel-junctions in which the barrier was semiconducting have been used by Giaever and Zellor.[86] The fabrication of this type of junction follows that of a tunnel-junction, with the substitution of a semiconductor such as cadmium-sulphide ~ 5 nm thick for the 1 to 2 nm thick oxide.

(ii) THE WEAK-LINK
The thin film weak-link type of junction, used by Anderson and Dayem,[54] *et al.*, is formed by evaporating the superconductor through a mask so that the deposited strip has a narrow neck, about 10^{-6} m square, at its centre. The film is about 10^{-5} m thick, and the formation of the narrow neck is rendered difficult by penumbra and diffraction effects at the edge of the mask. The mask has therefore comprised a "hair" of plastic cement stuck to the glass substrate and cut with a razor blade, the cut forming the neck. Alternatively microscribing has been used to form the neck.

(iii) THE POINT-CONTACT
The point-contact has been used extensively to form Josephson junctions. It has the advantage that, at varying contact pressures, the point penetrates the oxide on the plane electrode to varying extents. The coupling between the two superconductors may be varied and a range of current–voltage characteristics between the tunnel and weak-link type can be obtained. Both quasi-particle and

superconductive tunnelling may be observed. This type of junction is very sensitive to mechanical vibration, but the effects of vibration may be largely overcome by careful attention to the mechanical construction. Contaldo[87] and Kamper, *et al.*[60] have given useful designs.

A variant of the point-contact is the type using crossed oxidised wires. The wires are usually kept in contact by a bead of epoxy resin which is formed around the crossing point. Clark[72] has used an arrangement comprising a number of tin or lead spheres, $\sim$1 mm diameter and in light pressure contact with one another, to form an array of point-contacts for far infra-red emission and detection experiments.

(iv) THE SOLDER-DROP JUNCTION

The solder-drop type of junction was first used by Clarke.[77] It comprises a length of oxidised niobium wire which has a solder-drop 1 to 5 mm long formed around the centre. Two copper leads embedded in the solder-drop, together with the ends of the niobium wire, form a convenient four-terminal device (two current and two potential leads). The junction is formed between the niobium wire and the solder drop. Many of the commonly used solder-alloys are superconducting at about 7 K, e.g. see Warren and Bader.[88] The position at which the junction is formed is frequently controlled by making notches in the oxide with a razor blade. Single, double or multiple junctions may be formed, depending on the length of wire immersed in the solder-drop.

The solder-drop junction is essentially a dry-soldered joint having a resistance of about an ohm at room temperature (even a "dry" joint has its uses!). Experimentally it is found that about half of the drop-junctions having the required four-terminal resistance at room temperature form Josephson junctions when they are superconducting. This type of junction is therefore the most easily fabricated.

(v) DOUBLE JUNCTIONS

Double junctions have been formed using tunnel and point-contact junctions, and also with the solder-drop type. The two latter have been the most frequently used, and have found application as magnetometers and superconducting galvanometers[77] respectively. Figures 20(a) and (b) illustrate the tunnel and point-contact types respectively. The solder-drop type has been shown in Fig. 19(d) and may be either a single or a double junction depending on the size of the solder-drop.

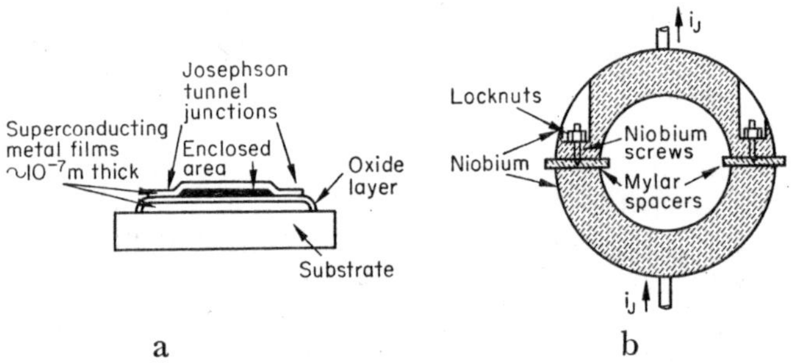

Fig. 20 Double junctions used as magnetometers; (a) *double tunnel junction and* (b) *double point contact, both used by Jaklevic, Lambe, Zimmerman and Silver. The solder-drop junction is also commonly used as a double junction.*

6. The Josephson Effects in Liquid Helium II

The possibilities for observing Josephson effects in liquid helium II were discussed by Anderson.[12] He concluded that it would be extremely difficult to realise the helium analogue of the superconducting tunnel-junction, for the barrier thickness would have to have sub-atomic dimensions owing to the much larger mass of the helium atom. However, the weak-link analogue did appear to be observable, and this conclusion led to the observation of Josephson effects in superfluid helium by Richards and Anderson[13] and later by Khorana and Chandrasekhar[14] and Khorana.[15]

The Josephson effects may be described in terms of the flow of quantised vortices; in superconductivity, they are the circulating current containing one flux quantum, but in superfluid helium II the vortices contain one quantum of angular momentum. In the liquid helium II experiments the vortices were generated by a quartz-transducer vibrating at frequency ν, placed just below the orifice, the vortex flow across the orifice representing the phase slippage. In the weak-link analogue the chemical potential difference is supplied by gravitational energy through the difference in level of two separate samples of superfluid helium. The samples are connected only by a very fine orifice $\sim 10^{-5}$ m diameter.

As expected, it was observed that whenever the relation $\mu_1 - \mu_2 = mgz = nh\nu$ was satisfied (where z was the difference in levels, g the acceleration due to gravity, m the mass of the helium atom and n the ratio of two integers), the flow of liquid between the two containers could be arrested. The transducer vibrations were used either to interrupt the flow to equilibrium of the levels, or to pump up the difference in levels.

The apparatus used by Khorana and Chandreskhar[14,15] is illustrated in Fig. 21. The inner beaker had a small orifice, 8×10^{-6} m diameter, in its thin sheet-nickel bottom (4×10^{-5} m thick). This orifice formed the only link between the two volumes of helium II. The top of the container was kept above the λ-point (2·2 K) to prevent the creep of the superfluid film linking the two liquid helium II samples. The inner beaker was made in the form of a coaxial capacitor so that the difference of 5·5 per cent between the permittivities of the helium vapour and liquid phases could be used to detect changes in the liquid helium II level as capacitance changes (0·11 pF/mm).

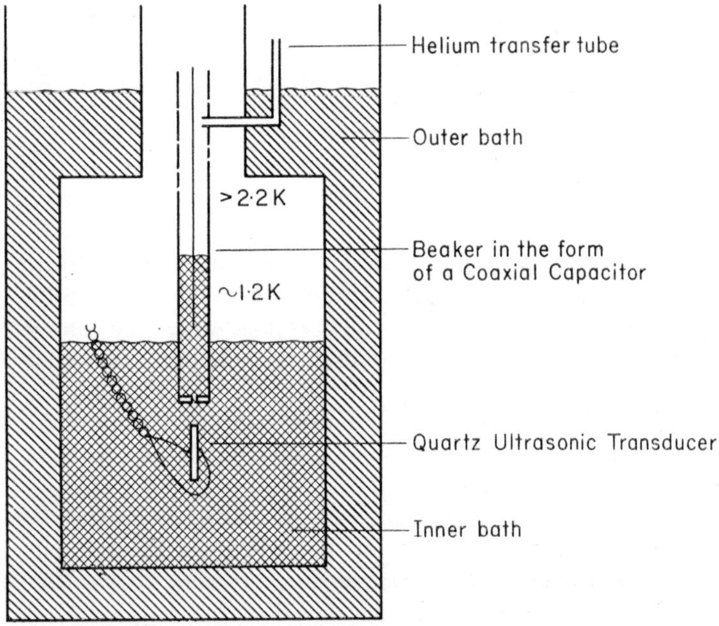

Fig. 21 The cryostat arrangement used by Khorana and Chandrasekhar[14] (1967) to study the Josephson effect in superfuid liquid helium.

Some of the results given by Khorana[(15)] are shown in Fig. 22. They show the pumping action of the transducer and the very clear steps in the levels. The steps could be maintained for long periods and were very stable, so much so that it was necessary to pulse the transducer power to move to another step. The changes in level for a transducer frequency of about 100 kHz were 1·01 mm. In principle then, the Josephson effects in liquid helium II enable one to see macroscopic quantum ordering effects with the naked eye!

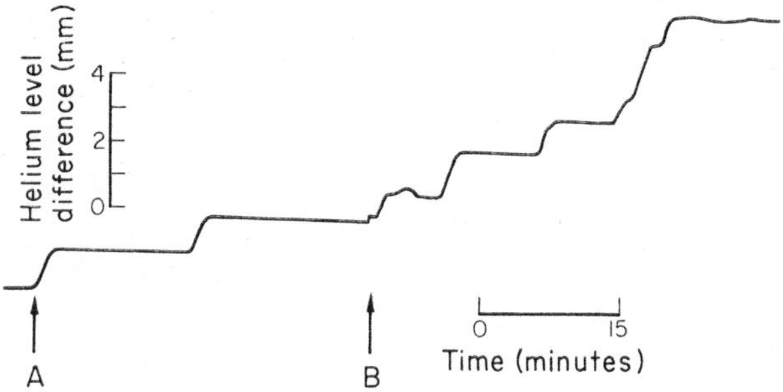

Fig. 22 The Josephson effect in superfluid helium II. Example of the variation of the difference between the helium levels in the two beakers, as a function of time, in the presence of vibrations from a transducer. At A *the transducer was switched on and at* B *the power increased. The transducer frequency was about 100 kHz and the corresponding steps at 1·01 mm intervals. The steps, z in the helium levels satisfy $mgz = nh\nu$, where m is the mass of the helium atom, g the acceleration due to gravity, n an integer or ratio of two integers and ν the frequency (after Khorana and Chandrasekhar,[(14)] 1967; Khorana,[(15)] 1969).*

It seems likely that further sophisticated experiments will be possible, such as the analogue of the double junction. However, even with the simple weak-line analogue, great care is necessary to exclude extraneous vortices (by, for example, careful cooling of the system through the λ-point, thermal shielding and ensuring freedom from extraneous vibrations); with the double weak-link these problems become much more pronounced.

7. Further Applications of the Josephson Effects

A. THE VALUE OF $h/2e$

The Josephson equation (16), $2\ eV = nh\nu$, is very simple and it is natural to wonder how precisely it is obeyed. From the theoretical standpoint it appears that although suggestions have been made that the emitted radiation may be pulled from that expected from the Josephson equation, for example by Stephen(89) and by Scully and Lee,(90) providing that the potential difference is measured as the electrostatic potential difference, as opposed to the electrochemical potential difference, the above equation is obeyed exactly. The frequency-pulling effects have been discussed by McCumber. (66)

Despite the elegance of theoretical arguments, it is always satisfying to test predictions experimentally. The group at the University of Pennsylvania therefore embarked on a series of classic experiments which led to their obtaining a value of $h/2e$ which had a standard deviation of 2·4 p.p.m. when referred to the National Bureau of Standard volt. The precision was slightly poorer when referred to the absolute volt of the SI definition of the basic units. In order to compare their determination (Parker, *et al.*(49)) with the value of $h/2e$ derived from the rest of physics they were obliged to undertake a re-evaluation of the fundamental constants of Physics (Taylor, *et al.*(91)), which evaluation they performed with skill.

The reason for their evaluation was that in the 1963 evaluation of the fundamental constants (Cohen and DuMond(92)), there was a difficult choice between two possible values of the fine structure constant occasioned by

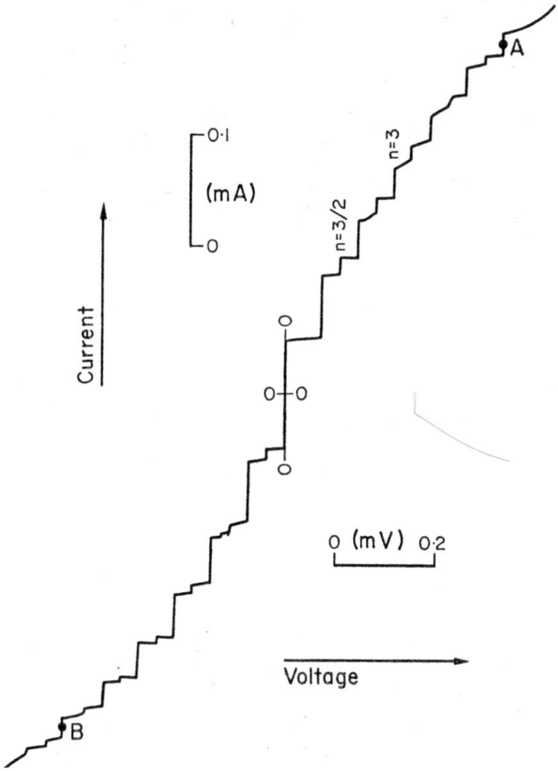

Fig. 23 An I–V characteristic of a solder-drop junction with 27 GHz radiation from a Gunn effect diode incident on it. Sub-harmonic steps are clearly visible. For 2e/h work a 36 GHz klystron was used and thermal e.m.f.'s eliminated by measuring the potential of corresponding steps on either side of the origin. The current was adjusted to the step centre as at A *and* B *(K. Morris and the author).*

discrepant sets of experimental data. DuMond and Cohen[93] suggested that further experimental work had indicated that the value adopted for the fine structure constant in their 1963–65 evaluation was probably incorrect. This conclusion resulted in a change of about 34 p.p.m. between the 1965 and 1969 values for the value of $h/2e$ deduced from the rest of physics. Taylor *et al.*[91] concluded that $h/2e$ by the Josephson effect agreed with the rest of physics to $2{\cdot}5 \pm 6{\cdot}4$ p.p.m. Some of their recommended values are shown in Table 1.

QUANTITY	UNITS	VALUE, TAYLOR *et al.* (1969)	ERROR (p.p.m.)
α^{-1}	—	137·036 02	1·5
e	10^{-19} C	1·602 191 7	4·4
h	10^{-34} Jsec	6·626 196	7·6
m_e	10^{-31} kg	9·109 558	6·0
N	10^{26} kmole^{-1}	6·022 169	6·6

Table 1. Recommended values for α^{-1}, e, h, m_e and N obtained by Taylor, Parker and Langenberg[91] *(1969)*

In their work (Parker, *et al.*[49]) they verified that the value of $h/2e$ was the same for emitted or incident radiation, did not depend on the temperature, material, step number, frequency, or whether a tunnel or weak-link type of junction was employed. Their result was confirmed by Petley and Morris,[94,95] using solder-drop junctions. The two results* agreed to well within a standard deviation ($\pm 2{\cdot}2$ p.p.m.). These precisions quoted for $2e/h$ are relative to the maintained national voltage standards, since the volt is only known absolutely to $\sim$3 p.p.m. More recently the Pennsylvania group have reported a value of $2e/h$ having a precision of 0·46 p.p.m. when referred to the N.B.S. volt (Finnegan, *et al.*[96]). The

* Proof note: Other workers [101], [112], [113] have now reported values which are also in excellent agreement.

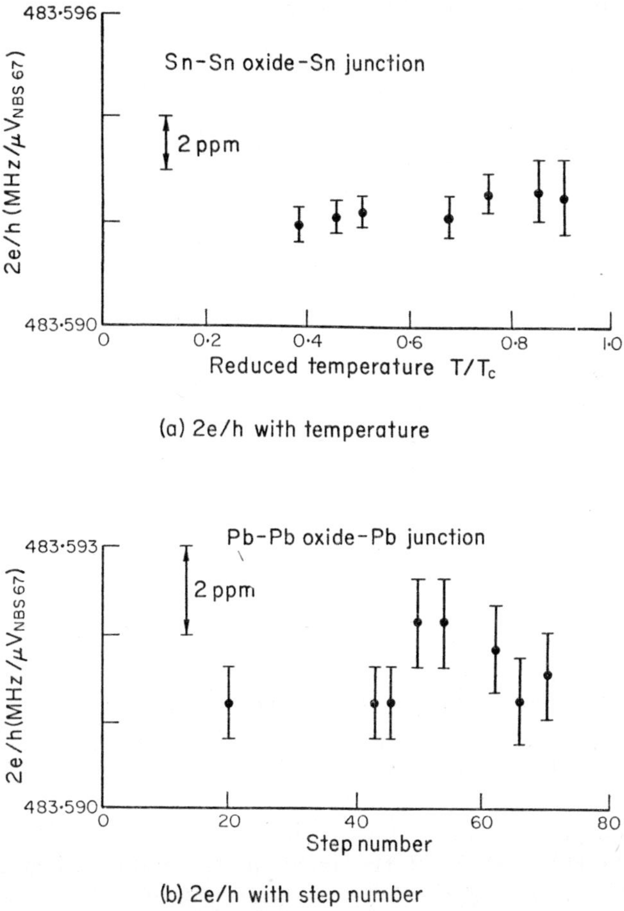

Fig. 24 The 2e/h results of Parker, et al.[(49)] *(1969),* (a) *for different temperatures and* (b) *for measurements on different steps. They are consistent with no correlation with either.*

sub-p.p.m. precision now achieved means that e/h may be regarded as an auxiliary constant in the future evaluations of the fundamental constants of physics.

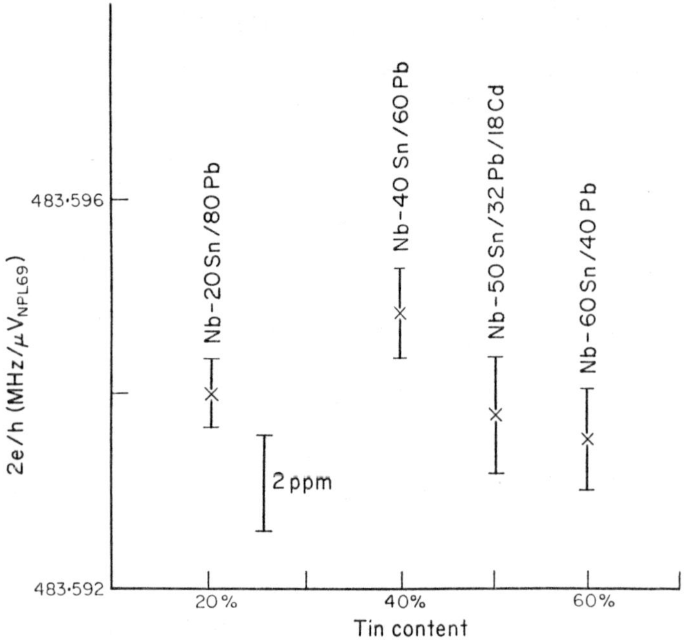

Fig. 25 The results of Petley and Morris (1969, 1970)[95] *for measurements of 2e/h on solder drop junctions with different solder alloys; they are consistent with no significant variation with alloy.*

Finnegan, *et al.*[97] found similar values for $h/2e$ for 9·2 GHz and 891 GHz incident radiation, using an HCN laser for the latter frequency source; the small uncertainty in its frequency limited the precision of their comparison to about 1·5 p.p.m. Further evidence of the precision with which the equation is obeyed was obtained by Clarke.[98] He measured the potential difference between two normal-metal-barrier superconductors, when biased on the same voltage step. Although the incident frequency range used by him was only between a few kilohertz and 1 MHz (the step voltage for the latter frequency being only about 2·1 nV), he was able to measure the potential difference between the two junctions very precisely using a solder-drop SLUG as galvanometer. In this way, Clarke verified that the junction potentials were the same to about a part in 10^8.

B. OTHER APPLICATIONS

Some of the device applications of the Josephson effect have been mentioned in the preceding sections, such as devices used as magnetometers or galvanometers, for $2e/h$ measurement, or possibly as a quantum voltage standard. The Josephson effects are already extremely competitive with existing devices for these applications. Many other device applications have also been investigated and reported in the literature. These applications have been reviewed,[(99)] for example, by Taylor.[(34)]

The radiation emitted by point-contact or tunnel-junctions enables them to be used as a simple source of millimeter wavelength radiation in situations where the complication of the helium cryostat is not a disadvantage. The point-contact is in many respects the more useful, because the frequency of the radiation emitted depends on the resonant frequency of the containing cavity; it is thus a broader band device. Silver and Zimmerman,[(100)] for example, used a Josephson junction as part of a simple radio-frequency spectrometer and with it they detected the nuclear magnetic resonance of Co^{60} in ferromagnetic cobalt.

The high degree of non-linearity of the effects in the barrier was utilised by Shapiro[(101)] to generate 12 GHz radiation from incident 4 or 6 GHz radiation. He used a solder-drop junction for this purpose. Harmonic frequency-mixing in a junction has also been demonstrated, for example by Grimes and Shapiro[(102)] at 23 GHz and 72 GHz. This harmonic-mixing property is a disadvantage in many applications, for the frequency response of a barrier ranges uniquely from d.c. to the infra-red and any extraneous radiation in this frequency range may be mixed at the barrier with the required signal. As a result, very well shielded cryostats and screened rooms are being used increasingly for definitive experiments with the Josephson effects.

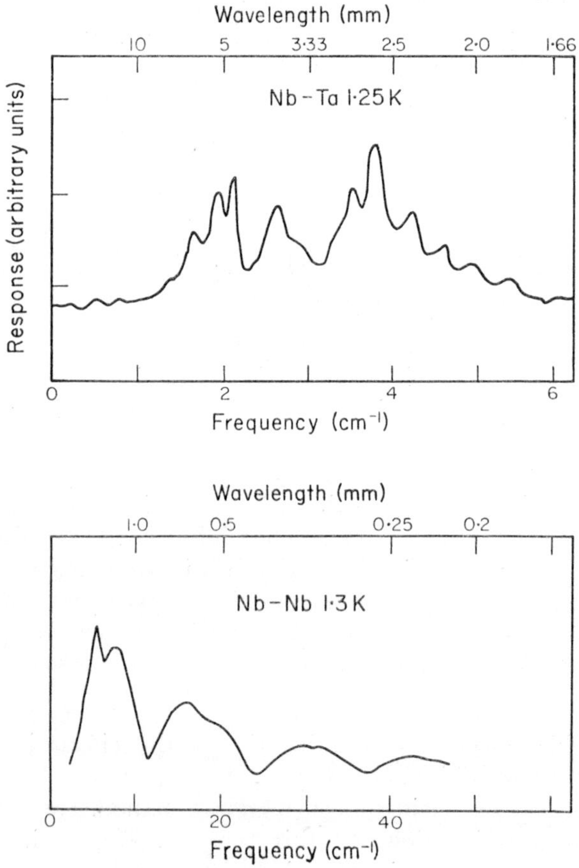

Fig. 26 Example of the response of a point-contact Josephson junction to infra-red radiation (after Shapiro,[105] *1967).*

The junctions clearly have possible uses as radiation detectors, and Grimes, Richards and Shapiro(103) have shown that the frequency response extends into the far infra-red to wavelengths shorter than ~300 microns with a very rapid response time of ~10 nsec and a detection sensitivity of less than 5×10^{-13} W. The response of two point-contact junctions to far infra-red radiation observed by Shapiro(105) is shown in Fig. 26.

The very rapid switching time(106) suggests applications of the junctions as switching elements for computers. Matisoo(85) found switching times of less than 10^{-9} sec with the type illustrated in Fig. 27. Schroen(107) has formed over 50 junctions on a single substrate. Since the resistance of the junction is negligible, and the switching time depends on the ratio of the inductance to the resistance of the junction, the junction and lead inductances must be maintained at a low level for fast switching applications.

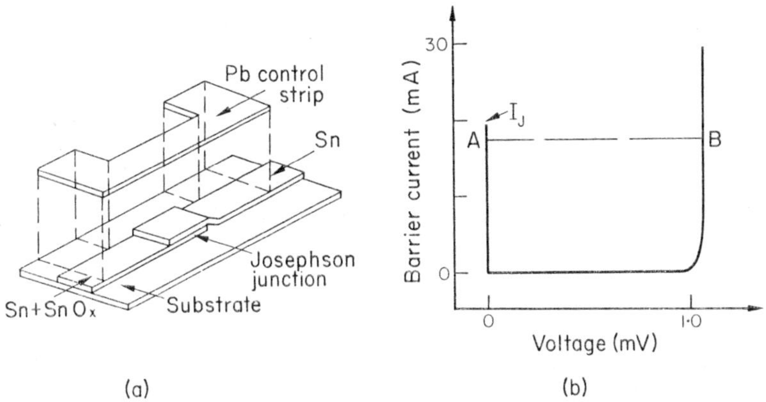

Fig. 27 The Josephson junction as a computer switching element. (a) *A parallel strip junction with the addition of a control-strip of lead;* (b) *the junction is biased to the point* A *just less than* I_J (*the amount depending on circuit noise*)*; a current through the control strip gives a magnetic flux that reduces* I_J *to less than that at* A *and the junction switches in less than* 10^{-9} *sec to* B (*after Matisoo,*(85) *1967*).

The applications of devices using Josephson effects depend on the amount of effort that is devoted to their development. They are rather fragile for applications where they must withstand frequent thermal-cycling to room temperatures, vibrations or electrical overloads. Transistors were of course equally fragile in their early development. The advent of any solid-state or other device that was capable of producing liquid helium temperatures over the small volume occupied by a Josephson junction could, of course, change the application possibilities overnight.

8. Conclusions

Josephson showed a remarkable insight in his original paper, and the Josephson effects have now been amply confirmed by experiment. Josephson's original paper has led to the publication of more than 300 papers, and it is inevitable that only a few of them have been covered in any detail in this short monograph. The Josephson effects have provided a fascinating confirmation of the B.C.S. theory and vindicated London's concept that superconductivity is a macroscopic quantum phenomenon. They have had a profound effect on low temperature physics and seem likely to have a similar technological impact.

It is difficult to foresee the final applications of the Josephson effects, for they are being used in so many different fields. Junctions have been flown in balloons by Ulrich, *et al.*[108] to detect the background infra-red radiation in the Universe, and used as magnetometers to detect the magnetic field from human heartbeats ($\sim 3 \times 10^{-11} T$ peak signal) by Cohen *et al.*[76] They have been used as galvanometers, as part of the sensing system on other instruments such as gravity meters, and in microwave sources, detectors and harmonic mixers and generators.

The measurements of $2e/h$ are being made with a precision thought unlikely a decade ago, and a "quantum volt" now seems much less of a pipe-dream. Standards laboratories will soon be maintaining and perhaps disseminating the "volt", using the inverse of the $2e/h$ experiment. If the precision implied by Clarke's findings can be realised in a more absolute experiment, the SI system may well be changed within a few decades.

There are still questions to be solved; for example, how far beyond the energy-gap may the supercurrent steps be seen. Finnegan, *et al.* have made $2e/h$ measurements on a tunnel-junction at a potential difference of about 10 mV using the 500th supercurrent step when the junction was irradiated with 9·2 GHz microwaves; incidentally they have also operated several junctions connected in series. McDonald, *et al.*[(109)] have observed structure in the characteristic of a point-contact out to 17 mV, when radiation from an HCN maser was incident on it suggesting a frequency response out to at least 8·2 THz, while other workers are rumoured to have seen steps out to even higher potential differences. Even a point-contact junction has self-resonant frequencies beyond $\sim 10^{12}$ Hz, so that the high voltages for observing steps reported do not necessarily imply a broad-band infra-red response. The parallel-strip type of junction is likely to be investigated in greater detail and other ways of reducing the self-field effects will be sought. Less approximate solutions of the non-linear differential equations will be obtained, and these will bring further interesting discoveries.

Physics is normally presented to the newcomer in such a tidy and apparently complete fashion that many feel depressed by the thought that all the interesting phenomena have been discovered. Brian Josephson's predictions were made while he was a post-graduate student at Cambridge[(110)] and form part of his Ph.D. Thesis.[(17)] Few better recent examples could be quoted to excite the enquiring mind.

REFERENCES

1. Josephson, B. D., *Phys. Letters*, **1** (1962), 251.
2. Anderson, P. W., *Progress in Low Temperature Physics*, **5** (1967), 1 (Ed. C. J. Gorter), North-Holland, Amsterdam.
3. Anderson, P. W. and Rowell, J. M., *Phys. Rev. Letters*, **10** (1963), 230.
4. Rowell, J. M., *Phys. Rev. Letters*, **11** (1963), 200.
5. Shapiro, S., *Phys. Rev. Letters*, **11** (1963), 80.
6. Fiske, M. D., *Rev. Mod. Phys.*, **36** (1964), 221.
7. Yanson, I. R., Svistunov, V. M. and Dmitrenko, I. M., *Zh. Eksperim. i Teor. Fiz.*, **A7** (1964), 2091; Engl. Transl., *Sov. Phys. JETP*, **20** (1965), 1404.
8. Dahm, A. J., Denenstein, A., Finnigan, T. F., Langenberg, D. N. and Scalapino, D. J., *Phys. Rev. Letters*, **20** (1968), 859.
9. Richards, P. L., Shapiro, S. and Grimes, C. C., *Amer. J. of Phys.*, **36** (1968), 690.
10. Collings, P. J. and Gordon, J. E., *Amer. J. of Phys.*, **37** (1969), 293.
11. Petley, B. W. and Morris, K. L., *J. sci. Instrum.* (*J. Phys. E.*), Ser. 2, **2** (1969), 649.
12. Anderson, P. W., *Revs. Mod. Phys.*, **38** (1966), 298.
13. Richards, P. L. and Anderson, P. W., *Phys. Rev. Letters*, **14** (1965), 540.
14. Khorana, B. M. and Chandrasekhar, B. S., *Phys. Letters*, **18** (1967), 230.
15. Khorana, B. M., *Phys. Rev.*, **185** (1969), 299.
16. Josephson, B. D., *Rev. Mod. Phys.*, **36** (1964), 216.
17. Josephson, B. D., *Advn. Phys.*, **14** (1965), 419.
18. Josephson, B. D., *Quantum Fluids* (Ed. D. F. Brewer), North-Holland, Amsterdam (1966).
19. Anderson, P. W., *Lectures on the Many-body Problems*, Ravello, 1963, **2**, 115 (Ed. E. R. Caianello), Academic Press, London (1964).
20. Feynman, R. P., Leighton, R. B. and Sands, M., *The Feynman Lectures on Physics*, **3** (1965), Chap. 21, Addison-Wesley, Reading, Mass.
21. Bloch, F., *Phys. Rev. Letters*, **21** (1968), 1241.

22. Bardeen, J., Cooper, L. N. and Schrieffer, *Phys. Rev.*, **108** (1957), 1175.
23. Lynton, E. A., *Superconductivity*, Methuen, London (1964). Rose-Innes, A. C. and Rhoderick, E. H., *Introduction to Superconductivity*, Pergamon, London (1969).
24. Betts, D. S., *Contemp. Phys.*, **10** (1969), 241.
Ginzberg, V. L., *Contemp. Phys.*, **9** (1968), 355.
Kurti, N., *Contemp. Phys.*, **8** (1967), 21.
Petley, B. W., *Contemp. Phys.*, **10** (1969), 139.
Taylor, A. W. B., *Contemp. Phys.*, **9** (1968), 549.
25. Bardeen, J., *Physics Today*, **40** (1969), (October).
26. Parks, R. D. (Ed.), *Superconductivity*, Vols. **1** and **2**, Dekker, New York (1969).
27. London, *Superfluids*, **1** (1950).
28. Ginzburg, V. L. and Landau, L. D., *J. exp. teor. Phys.*, **20** (1950), 1064.
29. Roberts, B. W., *Progress in Cryogenics*, **4** (1964), 159.
30. Gasparovic, R. F., Taylor, B. N. and Eck, R. E., *Solid State Commu.*, **4** (1966), 59.
31. Swihart, J. C., Scalapino, D. J. and Wada, Y., *Phys. Rev. Letters*, **14** (1965), 106.
32. Deaver, B. S. and Fairbank, W. M., *Phys. Rev. Letters*, **7** (1961), 43.
33. Doll, R. and Nabauer, M., *Phys. Rev. Letters*, **7** (1961), 51.
34. Taylor, B. N., *J. Appl. Phys.*, **39** (1968), 2490.
35. Giaever, I., *Phys. Rev. Letters*, **5** (1960), 147.
36. Schreiffer, J. R. and Wilkins, J. W., *Phys. Rev. Letters*, **10** (1963), 17
37. Schreiffer, J. R., *Rev. Mod. Phys.*, **36** (1964), 200
38. Adkins, C. J., *Rev. Mod. Phys.*, **36** (1964), 211.
39. Ambegaokar, V. and Baratoff, A., *Phys. Rev. Letters*, **10** (1963), 486, and erratum: *Phys. Rev. Letters*, **11** (1963), 104.
40. Riedel, E., *Z. Naturforsch*, **19a** (1964), 1634.
41. Goldman, A. M. and Kreisman, P. J., *Phys. Rev.*, **164** (1967), 544.
42. Schroen, W. and Pritchard, J. P., *J. Appl. Phys.*, **40** (1969), 2118.

43. Gor'kov, L. P., *Zh. Ekspr. Teor. Fiz.*, **34** (1958), 735; Engl. Transl.: *Sov. Phys. JETP*, **7** (1958), 505.
44. Anderson, P. W., Werthamer, N. R. and Luttinger, J. M., *Phys. Rev.*, **138** (1965), A1157.
45. Ferrell, R. A. and Prange, R. E., *Phys. Rev. Letters*, **10** (1963), 479.
46. Yamashita, T. and Onodera, Y., *J. Appl. Phys.*, **38** (1967), 3523.
47. Owen, C. S. and Scalapino, D. J., *Phys. Rev.*, **164** (1967), 538.
48. Langenberg, D. N., Scalapino, D. J. and Taylor, B. N., *Proc. IEEE*, **54** (1966), 56.
49. Parker, W. H., Langenberg, D. N., Denenstein, A. and Taylor, B. N., *Phys. Rev.*, **177** (1969), 639.
50. Eck, R. E., University of Pennsylvania, Ph.D. University Microfilms Ann Arbor, Michigan (1966), 67–3065.
51. Chen, J. T., Finnigan, T. F. and Langenberg, D. N. (1970).
52. Giaever, I., *Phys. Rev. Letters*, **14** (1965), 904.
53. Langenberg, D. N., Scalapino, D. J., Taylor, B. N. and Eck, R. E., *Phys. Rev. Letters*, **15** (1965), 294.
54. Dayem, A. H. and Grimes, C. C., *Appl. Phys. Letters*, **9** (1966), 47.
55. Zimmerman, J. E., Cowen, J. A. and Silver, A. H., *Appl. Phys. Letters*, **9** (1966), 353.
56. Ambegaokar, V. and Halpern, B. I., *Phys. Rev. Letters*, **22** (1969), 1364.
57. Stephen, M. J., *Phys. Rev.*, **186** (1969), 393.
58. Dahm, A. J., Denenstein, A., Langenberg, D. N., Parker, W. H., Rogovin, D. and Scalapino, D. J., *Phys. Rev. Letters*, **22** (1969), 1416.
59. Silver, A. H., Zimmerman, J. E. and Kamper, R. A., *Appl. Phys. Letters*, **11** (1967), 209.
60. Kamper, R. A., Mullen, L. O. and Sullivan, D. B., *NBS Technical Note 381* (1969).
61. Langenberg, D. N., Parker, W. H. and Taylor, B. N., *Physics Letters*, **22** (1966), 259.
62. Shapiro, S., Janus, A. R. and Holly, S., *Rev. Mod. Phys.*, **36** (1964), 223.

63. Taylor, B. N., Parker, W. H., Langenberg, D. N. and Denenstein, A., *Metrolgia*, **3** (1967), 89.
64. Scott, A. C., *Phys. Letters*, **25A** (1967), 132.
65. Waldram, J. R., Pippard, A. B. and Clarke, J., *Phil. Trans. Roy. Soc. Lond.*, **268,** (1970), 265
66. McCumber, D. E., *Phys. Rev. Letters*, **23** (1969), 1228.
67. Werthamer, N. R. and Shapiro, S., *Phys. Rev.*, **164** (1967), 523.
68. Albegova, I. Kh., Borodai, B. I., Yanson, I. K. and Dmitrenko, I. M., *Zh. Tekni, Fiz.*, **39** (1969), 911; Engl. transl.: *Sov. Phys. Tech. Phys.*, **14** (1969), 681.
69. Parks, R. D. and Mochel, J. M., *Rev. Mod. Phys.*, **36** (1964), 284.
70. Anderson, P. W. and Dayem, A. H., *Phys. Rev. Letters*, **13** (1964), 195.
71. Zimmerman, J. E. and Silver, A. H., *Phys. Rev.*, **141** (1966), 367.
72. Clark, T. D., *Phys. Letters*, **27A** (1968), 585.
73. Jaklevic, R. C., Lambe, J., Silver, A. H. and Mercereau, J. E., *Phys. Rev. Letters*, **12** (1964), 159.
74. Goldman, A. M., Kreisman, P. J. and Scalapino, D. J., *Phys. Rev. Letters*, **15** (1965), 495.
75. Smith, T. I., *Phys. Rev. Letters*, **15** (1965), 527.
76. Cohen, D., Edelsack, E. A. and Zimmerman, J. E., *Appl. Phys. Letters*, **16** (1970), 278.
77. Clarke, J., *Phil. Mag.*, **13** (1966), 115.
78. Pippard, A. B. and Pullan, G. T., *Proc. Camb. Phil. Soc.*, **48** (1952), 188.
79. Templeton, I. M., *J. Sci. Instrum.*, **32** (1955), 314.
80. Ries, R. P. and Satterthwaite, C. B., *Rev. sci. Instrum.*, **38** (1967), 1203.
81. Jaklevic, R. C., Lambe, J., Mercereau, J. E. and Silver, A. H., *Phys. Rev. Letters*, **12** (1964), 159.
82. Jaklevic, R. C., Lambe, J., Mercereau, J. E. and Silver, A. H., *Phys. Rev.*, **140** (1965), A1628.
83. Zimmerman, J. E. and Mercereau, J. E., *Phys. Rev. Letters*, **14** (1965), 887.
84. Zimmer, H., *Appl. Phys. Letters*, **10** (1967), 193
85. Matisoo, J., *Proc. IEEE*, **55** (1967), 172.

86. Giaever, I. and Zeller, H. R., *J. Vac. Sci. T.*, **6** (1969), 502.
87. Contaldo, A., *Rev. Sci. Instrum.*, **38** (1967), 1543.
88. Warren, W. H. and Bader, W. G., *Rev. sci. Instrum.*, **40** (1969), 180.
89. Stephen, M. J., *Phys. Rev. Letters*, **21** (1968), 1639.
90. Sculley, M. O. and Lee, P. A., *Phys. Rev. Letters*, **22** (1969), 23.
91. Taylor, B. N., Parker, W. H. and Langenberg, D. N., *Rev. Mod. Phys.*, **41** (1969), 375.
92. Cohen, E. R. and DuMond, J. W. M., *Rev. Mod. Phys.*, **37** (1965), 537.
93. DuMond, J. W. M. and Cohen, E. R., *Z. Naturforsch.*, **21a** (1966), 70.
94. Petley, B. W. and Morris, K., *Phys. Letters*, **29A** (1969), 289.
95. Petley, B. W. and Morris, K., *Metrologia.*, **6** (1970), n.2, 46.
96. Finnigan, T. F., Denenstein, A. and Langenberg, D. N., *Phys. Rev. Letters*, **24** (1970), 738.
97. Finnigan, T. F., Denenstein, A., Langenberg, D. N., McMenamin, J. C., Novoseller, D. E. and Cheng, L., *Phys. Rev. Letters*, **21** (1969), 229.
98. Clarke, J., *Phys. Rev. Letters*, **21** (1968), 1566.
99. Kamper, R. A., *IEEE Device*, **ED 16** (1969), 840.
100. Silver, A. H. and Zimmerman, J., *Appl. Phys. Letters*, **10** (1967), 142.
101. Shapiro, S., *J. Appl. Phys.*, **38** (1967), 1879.
102. Grimes, C. C. and Shapiro, S., *Phys. Rev.*, **169** (1968), 397.
103. Grimes, C. C., Richards, P. L. and Shapiro, S., *J. Appl. Phys.*, **39** (1968), 3905.
104. Richards, P. L. and Sterling, S. A., *Appl. Phys. Letters*, **14** (1969), 394.
105. Shapiro, S., *Proceedings of the Symposium on the Physics of Superconducting Devices, University of Virginia, Charlottesville* (1967).
106. Stewart, W. C., *Appl. Phys. Letters*, **14** (1969), 392.
107. Schroen, W., *J. Appl. Phys.*, **39** (1968), 2671.
108. "Josephson Detectors make Astronomical Observations", *Physics Today*, **23** (1970), n. 4, 55.

109. McDonald, D. G., Kose, V. E., Evenson, K. M., Wells, J. S. and Cupp, J. D., *Appl. Phys. Letters*, **15** (1969), 121.
110. Anderson, P. W., *Physics Today*, **23** (1970), n. 11, 23.
111. Harvey, I. K., Macfarlane, J. C. and Frenkel, R. B., *Phys. Rev. Letters*, **25,** (1970), 853.
112. Petley, B. W. and Gallop, J. C., *Proc.* (*1970*) *ICPMFC Conf.*, NBS Special Publl. 343, (1971).
113. Ilose, V., Melchert, F., Fack, H. and Schrader, H. J., *PJB–Mitt.* **81,** (1971), 8.
114. Kulik, I. O. and Yanson, I. K., "Josephson Effect in Superconducting Tunnel Structures", Science Publishing House, Moscow (1970). (Written and published in Russian this book contains references to the extensive Russian theoretical work on the Josephson effects.)

INDEX

M & B MONOGRAPHS

ME/1	**Tribology** E D Hondros	0.263.51588.5
ME/2	**Electron Beam Welding** M J Fletcher	0.263.51707.1
ME/3	**Vacuum Brazing** M J Fletcher	0.263.51708.X
ME/4	**Stress Corrosion Failure** P Greenfield	0.263.51725.X
ME/5	**Glow Discharge Material Processing** R A Dugdale	0.263.51787.X
ME/6	**Diamond Grinding** J Burls	0.263.51791.8
CE/1	**Microbiological Corrosion** G H Booth	0.263.51584.2
CE/2	**Solvent Treatment of Coal** W S Wise	0.263.51706.3
CE/3	**The Fluidised Combustion of Coal** D G Skinner	0.263.51714.4
CE/4	**Carbonisation of Coal** J Gibson D H Gregory	0.263.51774.8
CE/5	**The Principles of Gas Extraction** P Paul W S Wise	0.263.51802.7
CE/6	**Metal-Air Batteries** D P Gregory	0.263.51712.8
EE/1	**Threshold Logic** S L Hurst	0.263.51587.7
EE/2	**The Magnetron Oscillator** E Kettlewell	0.263.51641.5
EE/3	**Material for the Gunn Effect** J W Orton	0.263.51773.X
EE/4	**Microcircuit Learning Computers** I Aleksander	0.263.51724.1
EE/5	**DC Conduction in Thin Films** J G Simmons	0.263.51786.1
EE/6	**Reading Machines** J A Weaver	0.263.51839.6